For all who strive to make
new product development better
and are receptive to different and innovative
ways to improve the process.

" *We can judge our progress by the courage of our questions and the depth of our answers, our willingness to embrace what is true rather than what feels good.* "

Carl Sagan

Contents

Preface

With over 45 years of combined experience in consulting and training companies on how to follow the traditional phased-and-gated process, we have observed the shortcomings of the status quo and identified how product development could be done differently. By applying these lessons, we developed an approach, which we call Exploratory Product Development® (ExPD), to help companies improve their processes and practices.

Much of this success results from a key advantage of ExPD: it enables project teams to adapt to internal and external factors, and unique product nuances. We call the process Exploratory Product Development based on the definition of exploratory learning,[1] which is outwardly focused and uses experimentation to discover new technologies, develop new competencies, and create new business models.

Our work led us to develop the Product Risk Framework® tool, which supports the ExPD process to help product development teams identify, manage, and track uncertainty and risk reduction. We received a 2018 National Science Foundation (NSF) STEM I-Corp grant sponsored by the University of Chicago, Polsky Center for Entrepreneurship and Innovation to help us determine product-market fit. As part of this effort, we interviewed over 80 leading innovators from multiple product development disciplines and various industries, discussing uncertainty and risk in product development.

We wrote this as an easy to follow user guide for product development practitioners. This user guide summarizes what we have learned about product development and how businesses can apply ExPD to improve their own processes.

The many sources of information we drew from include the results of the NSF interviews, pilots of ExPD conducted in organizations, extensive secondary research, and our combined years of experiencing both victories and flops in product development.

We also wish to acknowledge three very helpful contributors. Therese Graff kept us on track, edited, and made substantial contributions on the topics of process, project management, and risk reduction. John Reeves provided his insight on ideation and culture. Special thanks to Ken Wojcieszek, a very competent and patient book designer.

[1] Aelong Wei, Yaqun Yi, and Hai Guo, "Organizational Learning Ambidexterity, Strategic Flexibility, and New Product Development," Journal of Product Innovation Management 31 (2014): 832–834.

How to Read This User Guide

The Exploratory Product Development (ExPD) process has been designed to support product development teams that develop physical products and/or software in either enterprises or start-ups.

Enterprises

If you're part of an enterprise, we recommend reading all the chapters since this user guide primarily targets established companies with existing product lines. Start by reading Chapters 1 and 2 "The Case for ExPD" and "High-Level Overview of ExPD" to get a flavor of the approach.

Start-Ups

If you're in a start-up and have a single product for which you're seeking a product-market fit, we recommend that you read Chapters 1 and 2. In Chapter 3 (on strategy), pay particular attention to the business model and Product Tier.

Then dig deep into Chapters 5 through 7 (the Explore and Create Segment). These chapters will help you identify, evaluate, prioritize, and then track the reduction of the most impactful uncertainties and risks of your product. This will demonstrate to potential investors and venture capitalists that you're making progress toward a successful finished product with reduced uncertainty and risk.

Later, as you grow your business and add additional products, go back, and read the other chapters.

TURBA CORPORATION

Introducing the Case Study:
The Turba Corporation

This user guide employs the fictitious company "Turba Corporation" to demonstrate the principles of ExPD in a case study format.

Turba is a US consumer electronics company specializing in home health care. Founded in 1959 and headquartered in Beaverton, Oregon, it has 825 employees and generates annual revenue of $900 million.

Turba is creating a new personal emergency response system (PERS) to target the growing senior market. The traditional PERS is a pendant with an emergency button. The user wears the pendant, and if they press the button, the security monitoring center is alerted. As with a home security system, the monitoring center then calls the chosen contact person and, if needed, sends an emergency response unit.

Turba's competition includes the PERS providers that typically charge for an initial purchase of the device, followed by recurring monthly fees. They have typically earned themselves a bad reputation for charging hefty fees.

These practices have caused mistrust among caregivers, the primary decision-makers when buying these types of devices. In this product category and through its users and caregivers, Turba has identified a prime opportunity to restore and build this trust.

Turba also wants to introduce innovation into the traditional PERS product. Traditional providers have not kept up with the physically active and tech-savvy senior. Turba seeks to extend the conventional PERS offering to a mobile version that allows users to travel anywhere at any time without being confined to the limited reach of a home-based system.

The Turba case study begins in Chapter 3 "Why Strategy Matters for Product Development," where you will read how Turba begins the ExPD process by addressing a business model and strategies. The following chapters continue applying ExPD principles in the accompanying Turba case studies. As an aid to identifying these case study examples, look out for the Turba Corporation logo.

Part I

Overview of ExPD

Chapter 1

The Case for Exploratory PD

Chapter 1 Contents

What to Expect

This chapter will build a case for an alternative approach to developing products: Exploratory PD® (ExPD). ExPD is an adaptable strategy-to-launch product development process that adjusts to internal and external factors, and unique product nuances. We begin with a discussion of uncertainty and risk, which are key considerations in product development. Then, to illustrate the value of adaptability, we offer an analogy: a comparison of Waze® navigation software and a static roadmap.

We identify some significant challenges companies face using the traditional phased-and-gated product development process and the trend toward more adaptable approaches to developing products, including a high-level recap of the different methodologies and processes some organizations use. We then wrap up the chapter with the Manhattan Project as an example of operating in an unstable, uncertain environment.

Uncertainty & Risk in Product Development

Decision-making in the face of uncertainty is a defining characteristic of product development. In this context, uncertainty is defined as imperfect or incomplete knowledge that results in the potential for surprise or unpredictability.

When, at time zero, you decide to proceed with a project, you're accepting the risk of an outcome resulting in either success or failure (see Figure 1.1). At six months, the product is launched. You know the actual first-year gain or loss at twelve months and can determine whether the project's outcome is successful or unsuccessful.

Figure 1.1: Decision-Making Under Uncertainty

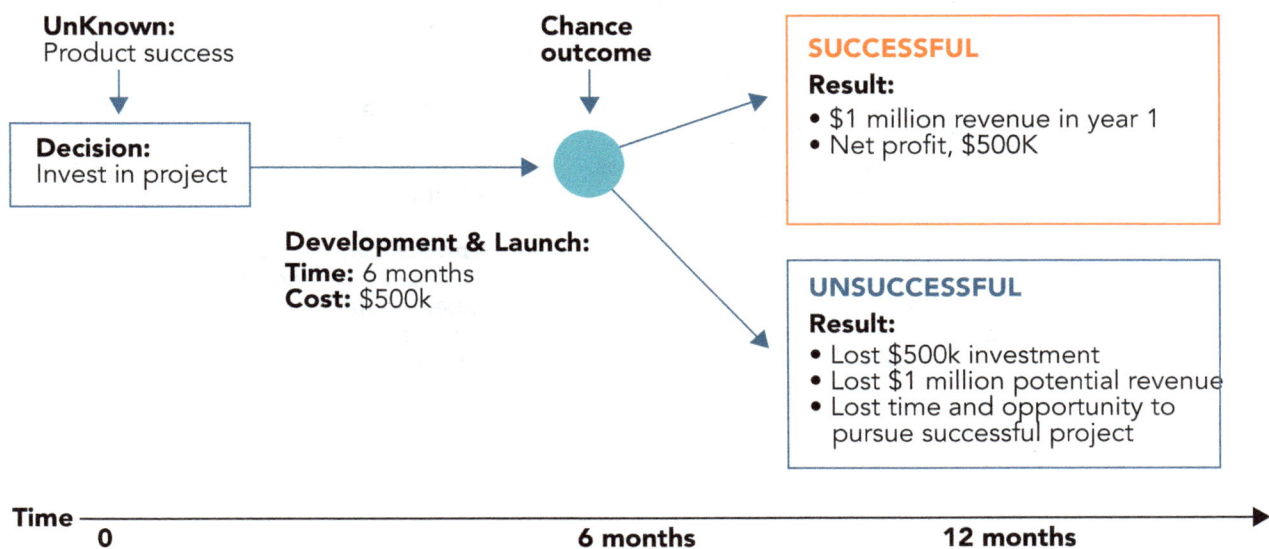

UnKnown:
Product success

Decision:
Invest in project

Development & Launch:
Time: 6 months
Cost: $500k

Chance outcome

SUCCESSFUL
Result:
- $1 million revenue in year 1
- Net profit, $500K

UNSUCCESSFUL
Result:
- Lost $500k investment
- Lost $1 million potential revenue
- Lost time and opportunity to pursue successful project

Time
0   6 months   12 months

Uncertainty is defined as imperfect or incomplete knowledge that results in the potential for surprise or unpredictability.

A primary goal of ExPD is to reduce the product uncertainty that derives from unknowns. In some instances, gathering the information to resolve the uncertainty is prohibitively expensive. It is also worth noting that unknowns may derive from false knowledge and mistaken beliefs.[1] As a result, risk reduction requires several possible actions (see Table 1.1).

Table 1.1: State of Knowledge

Actual state of knowledge:	What we believe our state to be: KNOWN	What we believe our state to be: UNKNOWN
Known	**Known Knowns** **Facts:** We have complete knowledge and don't need to seek out additional information. **Application:** Monitor these assumptions for changes in status.	**Unknown Knowns** **Blind spots:** We have prejudices, biases, and experiences that affect how we perceive reality and make decisions. **Application:** We need to identify these blind spots and move them into the category of known knowns.
Unknown	**Known Unknowns** **Uncertainties we are aware of:** The uncertainty may be due to unpredictable future events, such as losing key people, actions by competitors, or tasks taking longer than expected. We have standard ways of dealing with these uncertainties. Alternatively, the uncertainty may be due to a lack of knowledge or experience. **Application:** Experimentation designed to test hypotheses can move a known unknown to a known known.	**Unknown Unknowns** **We don't know what we don't know:** These are impossible to anticipate or even imagine based on our current state of knowledge. If they are discovered, it happens by surprise. They are rife in chaotic, changing and complex environments. **Application:** We can try to uncover unknown unknowns through experimentation but should be prepared to respond to surprises. Agility and adaptability are essential skills.

The ExPD process is designed to help identify and reduce uncertainties that can result in risk and product failures. But first, you must be able to recognize what uncertainty looks like and know where it lurks. ExPD addresses risks related specifically to product development, not just project risks, which are the uncertainties and risks encountered in executing a specific project, generally involving schedule, budget, and scope.

In everyday conversation, we tend to think of "uncertainty and risk" as something negative and to be avoided. Such generalities can get confusing, so we provide a cheat sheet of terminology with examples in Sidebar 1.1.

We found that our pilot teams respond better to a positive spin, stating uncertainties as assumptions rather than risks. An assumption is a statement that defines what the team needs to prove, or disprove, and is the basis for forming a testable hypothesis in an experiment. To provide a basis for logical reasoning or decision-making, we treat the assumption as true or certain to happen.

Risk may be positive or negative in impact. If there is no meaningful impact, then there is no risk.

Sidebar 1.1: The Language of Uncertainty and Risk

Uncertainty - The state of having imperfect or limited knowledge, results in the potential for surprise or unpredictability.

Example: Preliminary research suggests customers are willing to pay around $100 for our product. However, the actual amount they are willing to pay could be anywhere from $40 to $120.

Assumption - A statement that defines what the team needs to prove, or disprove, and is the basis for forming a testable hypothesis in an experiment. To provide a basis for logical reasoning or decision-making, we treat the assumption as true or certain to happen.

Example: We assume customers will be willing to pay $100. From this, we can reason: At a unit cost of $70, this price will yield a profit of $30 per unit. At a unit cost of $60, the profit will be $40.

Hypothesis - A proposed assumption that is tested in an experiment.

Example: We will test our hypothesis that customers are willing to pay at least $100.

Risk - A state of uncertainty that can result in either a positive or negative outcome. If there is no meaningful impact, then there is no risk. Sometimes, the term risk is used to describe only "negative" outcomes and not "positive" outcomes. We will use risk to cover both types of outcomes.

Example: We know our unit cost is $99, but we don't know the maximum price a customer is willing to pay. If it is more than $99, we can cover our costs and profit (positive impact). If the maximum price is $99 or less, we will lose money (negative impact).

Unknown - An unknown is something we don't know enough about. We may learn more and resolve the uncertainty, but this may prove impracticable or prohibitively expensive.

Example: We don't know if new competitors will enter the market. We could send out scouts worldwide to look for signs of potential competition, but it is prohibitively expensive.

Resolving Risk - Reducing the underlying uncertainty to an acceptable level. Rarely can all uncertainty be eliminated and sometimes product developers cannot reduce uncertainty, as in the case of competitor activity. If the organization accepts the risk, then it has resolved the risk.

Example: Based on our market research we are confident the customer is willing to pay at least $100. We accept this uncertainty as resolved.

For more risk terminology go to Appendix 1A
(The Language of Uncertainty and Risk)

The Need for Change in Product Development

Product development is important. Half of U.S. annual Gross Domestic Product is attributed to product and service innovation.[2] Further, in a survey of executives, 90 percent indicated the importance of developing and launching new products.[3]

Despite recognizing the importance of product development, most enterprises have difficulty developing products effectively. Typical pressures and frustrations include missed launch dates, missed sales goals, internal conflict, budget overruns, and the wrong products released to the marketplace. Often, the most common stressor is a sense of uncertainty.

As consultants, we implemented the traditional best-practice product development phased-and-gated processes (see Sidebar 1.2) for medium to large enterprises. As a result, companies saw improvements, but we still heard common complaints from our clients: burdensome documentation, glacially slow progress, and creativity-killing rigidity.

Sidebar 1.2. History of the Phased-and-Gated Process

Phased-and-gated processes became popular in the 1960s when cost containment and efficiency became management priorities.[4] Adoption by the U.S. Department of Defense and NASA helped establish the approach as a standard management best practice. Following its widespread acceptance in product development, it has become variously known as phased-and-gated, phased-review, tollgate, and PACE®, and in its most prominent and widely adopted iteration, as Stage-Gate®, introduced by Dr. Robert Cooper in 1985.[5,6] Figure 1.2 provides a high-level view of the Stage-Gate process.

Figure 1.2. The Full Stage-Gate Process®

0 Discovery	1 Scoping	2 Business Case	3 Development	4 V&V	5 Launch

Throughout this user guide, we will refer to this process as phased-and-gated; these processes generally have the same activities and process characteristics. Each phase (rectangle) comprises "a set of prescribed activities" where specific information is gathered, analyzed, and interpreted, from which prescribed deliverables are then produced.[7]

The phased-and-gated process is relatively linear and incorporates as few as three phases and as many as thirteen.[8] Every step of the process has prescribed detail: when activities should be done, who is responsible, what deliverables should look like, and how decisions are made.

This level of prescriptiveness seeks to reduce the risk of failure and loss from poor project implementation and decision-making. Furthermore, each phase gathers increasingly detailed information, which is also increasingly expensive to obtain.

Depending on product type, there may be different process paths: a comprehensive process for new or complex products, a consolidated process for simple revisions.

At the end of each phase, the gate (diamond) provides an opportunity to judge the project and decide on continued investment, place it on hold, recycle, or cancel. Each gate also facilitates management oversight of the process, with incremental step-wise commitments to investment and risk exposure.[9]

For us, the straw that broke the camel's back came in the form of a very dissatisfied R&D director at a Fortune 500 company where we assisted in designing and implementing a phased-and-gated process. While the director accepted that the approach would provide structure, he was particularly resistant to its adoption because it would be too slow and unable to adapt to his business's ebbs and flows.

But if the industry-standard best practice—the phased-and-gated approach—wasn't the best methodology for his company, what was? At the time, there were no good alternatives.

But if the industry-standard best practice—the phased-and-gated approach—wasn't the best methodology for his company, what was?

We are very familiar with the traditional phased-and-gated methodology, having 40-plus years of combined experience consulting and training companies on how to use it. However, by witnessing the problems associated with the phased-and-gated process, we saw opportunities to improve the product development process. Our solution is ExPD, a systems approach that fully integrates all the necessary elements and offers a fundamental redesign of the development process.

The best way to illustrate the differences between ExPD and the phased-and-gated process is with an analogy. Think about the disadvantages of relying on the tried-and-true roadmap. Generations of travelers used printed road atlases or maps to plan and manage their trips. They usually worked, but the best available routes could be out of date by the time they were printed, and travelers would need to resolve unexpected construction or heavy traffic along the way.

Today, GPS technology apps such as Waze® are adaptive. Based on real-time traffic and route information, alerting travelers to problems ahead and allowing route adjustments reduces overall delays and frustration (Figure 1.3).

Likewise, when navigating product development, the phased-and-gated process is static, whereas ExPD offers adaptability and flexibility in response to changing conditions.

ExPD offers the adaptability of a navigation app like Waze.

Figure 1.3: Travel Analogy ExPD

Traditional product development is to a road atlas ...

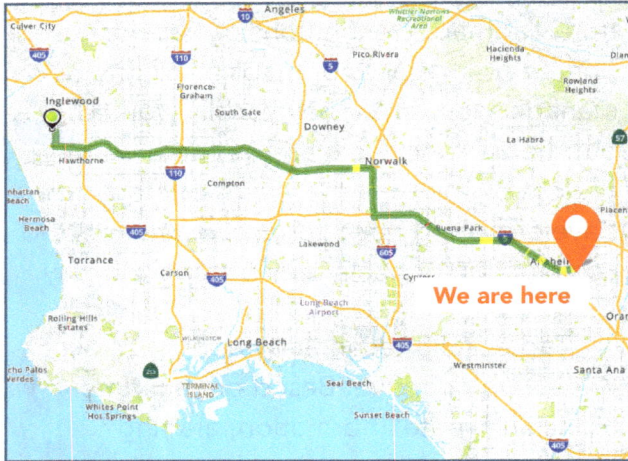

Early definition is like using a road atlas to plan and manage a trip. The route is predetermined, and travelers have little ability to modify the route if they encounter road construction or traffic along the way. Delays become a surprise and are difficult to deal with.

Source: Mapquest.com

. . . as ExPD is to Waze®

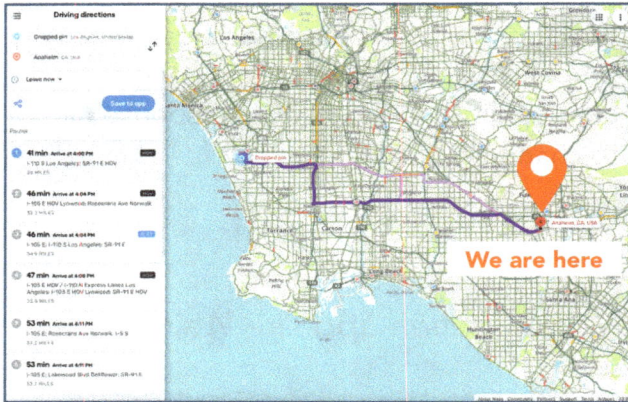

ExPD is an adaptive approach and it operates like Waze. It is an app that plans a driver's route based on experience and current conditions. Throughout the trip, it continuously monitors road conditions ahead, using real-time feedback. The process is adaptive and learns how to react to roadway surprises quickly.

Source: Waze.com

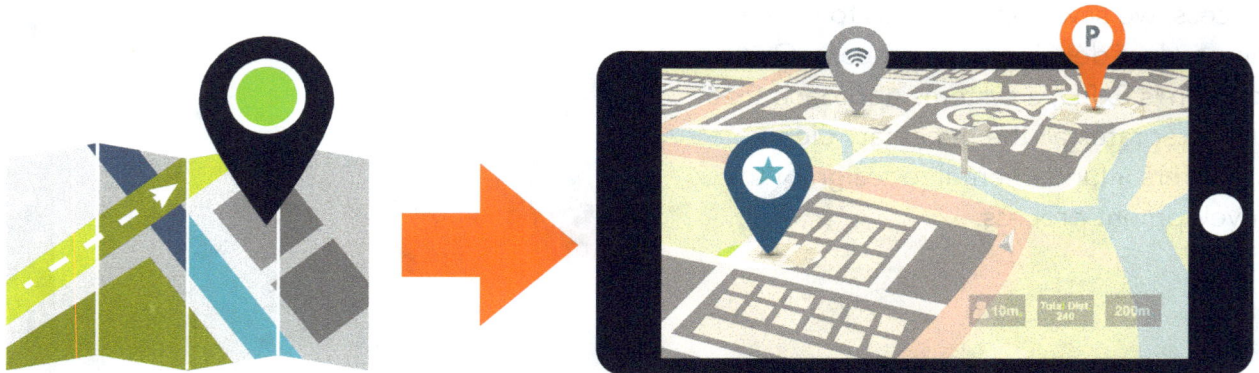

The Underlying Problem: Uncertainty

As discussed earlier in this chapter, the underlying problem with product development is uncertainty, which creates risk and the potential for loss (financial, time, opportunities, etc.). As product developers, we try to read the market correctly, but even if we are right today, how do we know if the market will remain stable? What if future competition renders our product obsolete? What if our product technology doesn't work the way we expect? What if future regulatory changes make our product more expensive to produce?

In product development, the underlying problem is uncertainty, which creates risk and the potential for loss.

Uncertainty coupled with the complexity of the external environment means that product development teams often make decisions based on insufficient information. For example, you may have a good understanding of the customer and the competition but relatively little knowledge of the technology. Alternatively, you might be confident in using a particular technology while knowing little about the market. In either case, a high level of uncertainty means a high risk of costly errors—not only in money and time but also in lost opportunities.

Academics have identified three main types of uncertainty related to product development: (1) market-related uncertainty, (2) technology-related uncertainty, and (3) project scope uncertainty.[10]

In one survey, more than half (58 percent) of project managers cited technology-related uncertainty as a reason for delays in product development.[11]

In one survey, more than half (58 percent) of project managers cited technology-related uncertainty as a reason for delays in product development.

Research has also shown that different levels of uncertainty require different approaches, and current product development processes, such as the phased-and-gated process, do not address the complexity of highly uncertain and risky products.[12] Another research study observed that the phased-and-gated process harms innovation because it cannot support complex revisions when uncertainty and risk are high.[13]

Phased-and-gated processes can harm innovation when risk and uncertainty are high.

Even though uncertainty is at the core of product development failure, we have found that most companies' product development processes do a poor job of identifying and managing it. In some instances, managers get involved at the micro-level, looking at every detail. Before making any decisions, they want regular reviews backed up by extensive analysis and documentation, slowing down the whole process and spending more time defending the project than working on it. Meanwhile, despite all this oversight, they're still not identifying or resolving the risks.

Too Slow for the Times

Companies can no longer afford to rest on their laurels but must be responsive to change and act fast. When the rest of the world is changing quickly due to globalization, increased competition, and new technologies, slowness carries its own set of risks. Missed opportunities and the release of outdated products cut into the bottom line. When you delay activities and decisions to specific points in the process, you risk wasting resources on projects that should be canceled.

Delay is intrinsic to the phased-and-gated approach, which seeks to define and specify every single step of the process with prescribed activities, and expected deliverables. Teams using this approach find that their timelines get longer and longer as they engage in unnecessary tasks. Also, the postponement of decisions to gate meetings results in wasted resources and slow adaptation to changing conditions. The phased-and-gated process results in large batches of activities, leading to delays (See Sidebar 1.3).

Sidebar 1.3. Avoid Large Batches to Gain Speed

In a phased-and-gated process, each phase is a batch of activities that must be completed to produce a set of deliverables. These deliverables are then reviewed in a batch at a gate. The problem with this method is that the batch is held until its longest activity is completed.[14] The larger the batch, the longer it will take to complete, and the greater likelihood the work will be delayed.

In contrast, ExPD emphasizes adaptability and flexibility, not large batches. As a result, the team gains speed by breaking up project uncertainties into smaller batches and short iterations. Several benefits can result:

- **Faster feedback.** Quick experimentation and iteration allow the team to quickly identify and resolve the most significant risks and uncertainties.

- **Faster cycle times.** Typically, each item in a batch cannot advance to the next step until the batch's longest activity is completed. In smaller batches, fewer activities are on hold, waiting for approval to move on.

- **Shorter queues.** We know from queuing theory that smaller batches result in less idle time.[15]

As companies pursue more innovative products and react to globalization, they must grapple with new uncertainties and risks, requiring new skills, knowledge, and resources. Processes need to adapt to the needs of each product. The phased-and-gated process hasn't changed substantially since its origins, more than 50 years ago. Companies have experimented with modifying the phased-and-gated process to add speed, reduce wasted activity, and incorporate more customer feedback.

Cooper describes these as elements of Next-Generation Stage-Gate: multiple processes for different risk levels, fuzzy gates, overlapping phases, adding market feedback "spirals."[16] Despite easing some process rigidity, the structure of the process is maintained. The nature and order of activities and deliverables are still prescribed, and decisions are deferred to the gates.

The phased-and-gated process hasn't changed substantially since its origins more than 50 years ago.

The ExPD Advantage

ExPD resolves the product's most significant uncertainties through experimentation and learning rather than following prescribed activities in a phased-and-gated process, thus "failing smarter and faster." As a result, the project team understands sooner what could potentially invalidate the project. Figure 1.4 illustrates how ExPD offers these advantages, using the hypothetical example of a wind turbine manufacturer.

Figure 1.4: Infographic Rotor Blade

Situation:

A project team within a large wind turbine manufacturer was asked by senior management if they could build a specialized rotor blade that was larger than they had ever built before.

Scenario 1: Exploratory PD (ExPD):

At the beginning of the project, the team identified and evaluated the biggest project risks and uncertainties. Shipping the blade was the biggest unknown.

In order to resolve the uncertainty about shipping the blade, the team built a prototype based on the expected dimensions and weight of the blade.

Approximate time:

2–3 weeks to build and test the shipping prototype

Scenario 2: Phased and Gated Process

0 Discovery	1 Scoping	2 Business Case	3 Development	4 V&V	5 Launch

The project team starts in the early phases to determine if the project is commercially and technically feasible. If the project appears to be feasible, it then proceeds to Development. Extensive paperwork is filled out within each phase by multiple departments. Operations then determines if the specialized blade can be shipped during the Verification & Validation (V&V) phase.

Approximate time:

3–4 years

What happens if they can't ship the blade?

In **scenario 1**, you lost **2-3 weeks** prior to a significant investment in the project.

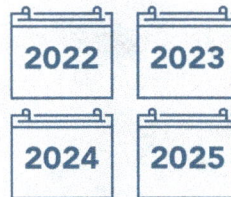

2022　2023

2024　2025

In **scenario 2**, you lost **3-4 years** with significant resource investment and lost opportunity cost.

Which scenario do you prefer?

In this example, the cross-functional team identified, evaluated, and prioritized the most impactful uncertainties and risks of the wind turbine project. In scenario 1, the team used ExPD and shifted the most impactful uncertainty of the project—shipping the blade—to front and center. The team worked toward resolving this important uncertainty before proceeding with the project. If the blade cannot ship in this scenario, the loss translates to two to three weeks without any significant future investment in the product. A decision needs to be made to either cancel the project or spend more time redesigning the blade.

In scenario 2, the traditional phased-and-gated process is rigid and prescribed activities do not always catch unique product nuances, like shipping the blade. Preparing for the blade shipment doesn't occur until the V&V phase. If the blade cannot be shipped, the loss translates to a whopping three to four years of resources and significant investment, not to mention the lost opportunity cost. The approach and outcome of these two different methodologies are very apparent. ExPD adapts to the product needs versus the process needs.

ExPD adapts to the product needs versus the process needs.

The Trend toward More Adaptable Processes

Products that require new skills, knowledge, experience, assets, and changes to the business model require a different product development process. Recent thinking in product development identifies shortcomings of the traditional approach and supports our observations.

According to the 2012 Comparative Performance Assessment Study conducted by the PDMA Research Foundation, there is a trend toward flexibility in product development. The study states, "The use of formal, structured processes may have reached its potential, and companies are now testing the use of more flexible methods."[17] Formal structured processes in large companies decreased from 72 percent in 2004 to 55 percent in 2012, favoring innovation that was more radical, innovative, and incremental.

"The use of formal, structured processes may have reached its potential, and companies are now testing the use of more flexible methods."
PDMA Research Foundation

Pricewaterhouse Coopers's 2014 Global Innovation 1000 study indicates that one of the most impactful expectations is a shift in R&D spending to support newer and breakthrough products.[18] It sounds simple, but this is a significant change for most organizations. It means taking on much riskier projects. That requires strong risk identification, measurement, and management to ensure that projects are investments and not a waste of resources.

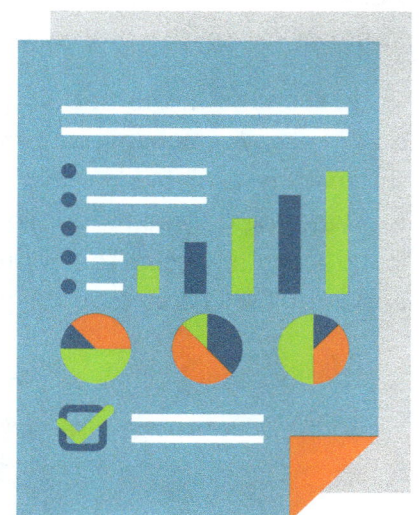

ExPD: Adaptable Strategy-to-Launch Process

Making the development process adaptable to the needs of each product is crucial. Some organizations have recognized this need and adopted various approaches and methodologies including Design Thinking, Agile Product Development, Flexible Product Development, Lean Startup and Lean Product Development. Unfortunately, none of these communities yielded an adaptable strategy-to-launch process for enterprises to follow, so we developed ExPD.

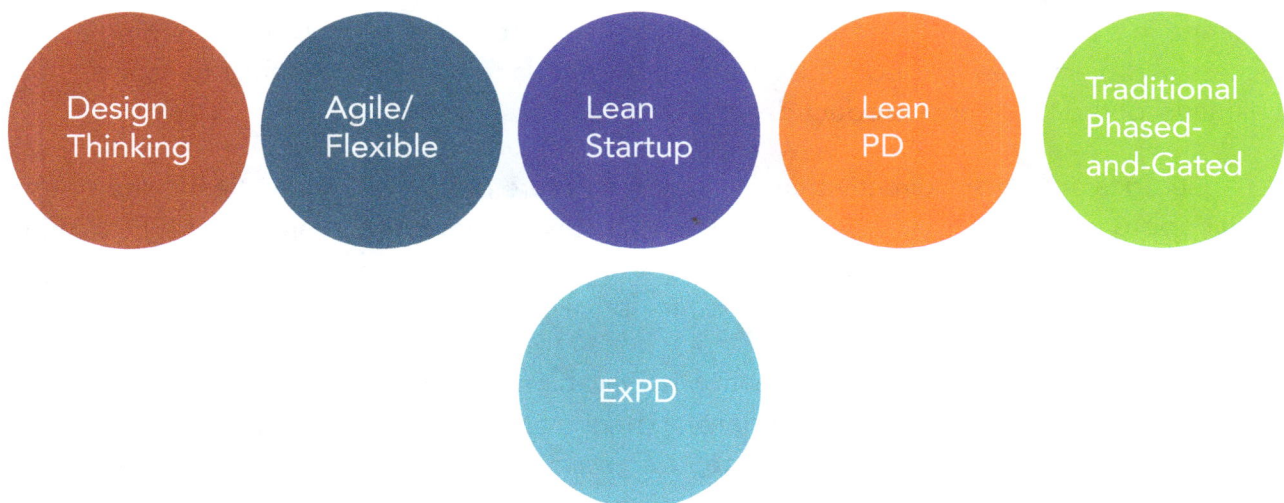

Design Thinking

Agile/ Flexible

Lean Startup

Lean PD

Traditional Phased- and-Gated

ExPD

The development of ExPD involved rigorous research into existing product development practices and processes. We leveraged some of these practices into our methodology, and we provide a brief overview of each community below.

The Design Community

Design Thinking is a process and toolset for designing and evaluating innovative solutions with the end-user, also known as Human-Centered Design or User-Centered Design. Design Thinking focuses on developing a deep understanding of end-users, including identifying a multi-dimensional problem, solving unmet needs, and generating creative solutions through methods like ethnography and brainstorming. Design thinking tools can be used throughout the process.

The Agile/Flexible Community

Agile is a software development approach that adapts to changing or emergent customer needs and changing technologies and environments. Software is developed in short iterations (one to four week intervals) with frequent software releases to customers to facilitate quick feedback and modification of the next iteration. This fast, iterative approach means that a change in customer requirements will not cause havoc to the project. Thus, managing a successful project does not require extensive up-front planning and adherence to scope (hallmarks of the traditional product development process).

Agile does not always directly transfer to non-software products. Preston Smith describes a similar approach for non-software products, called Flexible Product Development (FPD).[19] FPD allows changes to the product relatively late in development by incorporating specific design or architecture decisions without being too disruptive.

Delaying decisions until more information is available is one way of managing risk. It is valuable when elements of the product development environment—such as customer needs, regulations, technological advances, competitive activity, and organizational needs—are in flux. FPD is not a process but rather a set of tools and concepts for improving product development.

The Lean Startup® Community

Lean Startup (also referred to as Lean Launchpad®) grew out of the community of entrepreneurial internet-based software start-ups looking for venture capital backing. This approach eschews the traditional business plan based on assumptions around the customer, market, product, and business model on the basis that they are typically groundless. Instead, Lean Startup focuses on finding and demonstrating the product-market fit (the most significant risk for a startup) and creating a working business model through managed experimentation. A core concept of this approach identifies and resolves assumptions and risks through fast iteration and feedback.

The Lean Startup approach has been fine-tuned for internet-based startups. It has delivered impressive results partly because the tools and techniques rely on entrepreneurial company culture and early-adopter customer behaviors. Steve Blank, the founder of this community, has also applied these techniques to a program called Hack for Defense, a cross-collaboration between the Department of Defense and 40 colleges and universities.[20]

At an enterprise level, Lean Startup practices were adopted by GE (General Electric) through its FastWorks™ program. During the 2014 Lean Startup Conference in San Francisco, GE speaker, Cory Nelson, reported that only 15 to 20 percent of all employees understood the core elements of Lean Startup,[21] requiring a mind shift in most of the thousands of GE's employees being trained.

The Lean Product Development Community

Lean Product Development grew out of the application of lean manufacturing principles to reduce waste and improve speed in product development. These approaches are also referred to as the Toyota Development System (TDS), Design for Lean Production, and Product Development FLOW.[22]

The tools and techniques used in these lean approaches help address traditional product development systems' shortcomings: bureaucracy and delays.

The Traditional Product Development Community

We classify the "traditional" product development community as established companies that attempt to follow a predetermined process. Please go to Sidebar 1.2 if you wish to learn more about the traditional product development process. Table 1.2 provides further information on the differences between ExPD and the phased-and-gated process.

Table 1.2: Differences between ExPD and the Phased-and-Gated Process

System Element	ExPD	Phased-and-Gated Process
Process characteristics	The product drives the product development process Adaptable activities Activities and timing are based on the uncertainty and risk of the project. Iterative risk reduction is the primary principle behind the process and structure	The process drives the development of the product Sequential activities Activities and documentation are prescribed within each phase
Strategic framework	Strategy and roadmaps are tightly integrated into the product development process Strategy is adaptable; based on the enterprise's environment/market	Strategy and roadmaps are treated separately from the product development process
Resource Management	Prioritization and allocation of projects and resources are executed in real-time to prevent delays and overloading the pipeline	The assignment of resources is not directly incorporated within the product development process. Resource management may be addressed at regular intervals, such as at a portfolio management meetings
Risks	Risks drive the process Critical risks are pulled forward for resolution, enabling "learn fast and adapt" decisions Risks are explicitly identified, evaluated, and prioritized at the project's onset and revisited and updated during the entire process	Prescribed activities drive the process. Risk identification and the process generally start at the development phase when technology risks are high
Documentation	Modular, targeted documentation is reusable. Project teams document only the assumptions they are resolving	Documentation addresses all aspects of the project with increased detail at each phase
Decision-making	The team has clear guidelines on resources, budgets, and timelines Decision-making is decentralized, and management by exception is incorporated The project team can recommend adapting the project quickly at any time	Monthly gate meetings are typical. Gatekeepers make decisions (go, cancel, recycle, hold) Gatekeeper involvement depends upon the organization (management, director, or senior level)
Team skills	More experienced project leads typically lead the team Team members are open to ambiguity	Less experienced project leads can lead the team Team members require structure and a formula to follow
Product types	ExPD can work across all product types. It is especially beneficial for products with a high level of uncertainty ExPD can be used for hardware and/or software	Teams typically have different process paths, depending on product type: full process for new or complex products, a lite process for revisions It is not recommended for software products

Manhattan Project

In closing, we provide an overview of one of the most iconic projects in 20th-century history—the Manhattan Project. This project provides principles for operating in an unstable, uncertain environment.

The Manhattan Project had to produce an effective nuclear bomb in a very short time. The two critical priorities were time to completion and technical performance. The project met its goal in less than three years by embracing a scientific approach and a willingness to adapt to unforeseen events.[23]

By adopting a scientific approach and willingness to adapt, the Manhattan Project was completed in less than three years.

In a California Management Review article, Sylvain Lenfle and Christoph Loch explain how the concurrent pursuit of different solutions (set-based design) and uncertainty-reducing experimentation methods made the Manhattan Project attainable.[24]

The two most important uncertainties facing the Manhattan Project were the bomb design and the selection of fissile material. Very little was known about either, so the project activities and product definition had to be determined during the project's execution.

The team pursued several bomb designs and fissile material options at the same time. Building concurrent solutions proved to be much faster than a sequential approach.

The learning in the concurrent investigations proceeded through trial and error. Teams developed hypotheses and executed experiments. Then, based on the learnings, they updated their hypotheses and conducted more experiments until they determined whether the solution they were pursuing was viable.

Experimentation and concurrent pursuit of options enables the required speed-to-solution. Unfortunately, these concepts have been generally lost from the product developer's tool kit, even though they effectively reduce time-to-market.

Even though your projects may not be on the scale of the Manhattan Project, the practice of experimentation is key to reducing uncertainty and improving product development success. ExPD embraces these principles.

In the next chapter, we will provide a high-level overview of the ExPD process.

Key Chapter Points

1. The underlying challenge with product development is uncertainty, which creates risk and the potential for loss.

2. Uncertainty is the state of having imperfect knowledge or a lack of knowledge, resulting in surprise or unpredictability.

3. The ExPD process is designed to help identify and reduce uncertainties that can result in risk and product failures.

4. Phased-and-gated processes began to gain popularity in the 1960s as cost containment and efficiency became management priorities.

5. In product development, the phased-and-gated process is as static as a road atlas, while ExPD offers the adaptability of an online navigation app such as Waze.

6. More than half of project managers have cited technology uncertainty as a reason for product development delays.

7. Research has shown that while the phased-and-gated process is useful for simple revisions, it can harm innovation when product risk and uncertainty are high.

8. The phased-and-gated process tends to generate large batches, resulting in delays.

9. The ExPD process is adaptable to the needs and unique nuances of the product.

10. Many product development communities offer excellent practices and methodologies, but ExPD provides a unified, adaptable strategy-to-launch process.

11. The Manhattan Project demonstrated the importance of set-based design and experimentation.

Appendix 1A: The Language of Uncertainty and Risk

Known

Something is considered known when the level of uncertainty is acceptable.

Example: A company is planning to develop a new software product. The technology incorporates a recommendation algorithm in which the company has no experience— there is high uncertainty which can impact the success of the project. In comparison, they have extensive experience creating a web-based user interface and are highly confident they can build the entire user interface with no problems. The ability to build the user interface is considered known.

Fixed

A fact or model that will not change and acts as a constraint. In ExPD, the business model is usually considered fixed (unless the company is intentionally pursuing a new business model). Something that is fixed is also considered known.

Example: A company sells its products through a direct sales force and has no other channels. The sales channel is fixed.

Black Swan

A Black Swan is Nassim Nicholas Taleb's term for an event that meets three characteristics: (1) it happens rarely (perhaps it has never been observed before); (2) it has an extreme impact, which can be positive or negative in consequence; and (3) it is not predictable but in retrospect is easily explained.[25] This type of risk is a problem for us humans because we are hardwired to see the normal or the average.

As a result, we get blindsided by Black Swans with surprising regularity. Taleb's warning is that we need to be conscious of the potential for Black Swans in our risk identification processes.

Examples: Taleb's examples include the 9/11 terrorist attack, and the rise of Hitler and World War II.

Degree of Uncertainty

Measurement of uncertainty along a continuum. One end represents uncertainties that are resolved (made certain or close enough). The other end represents uncertainties, where obtaining the facts is impossible or too expensive.

Example: High uncertainty exists if a company sees an opportunity for a new product but has no experience with the market, the type of product, and the key technologies. Whether the company would be successful in developing and launching this product is highly uncertain.

Note: Often, the degree of uncertainty is subjectively rated on a scale of 1 (extreme uncertainty, meaning little to no information) to 5 (complete certainty, meaning almost complete to total information). The degree of uncertainty applies only to the likelihood of something occurring or being true; it is different from the Degree of Risk (see next page).

Degree of Risk

The degree of risk is a measurement of risk along a continuum. One end represents no possibility of loss or opportunity (either the probability of occurrence has been eliminated, or we have found a way to circumvent the issue should it occur). The other end represents risks that cannot be eliminated because the necessary information is impossible or too expensive to obtain.

Example: A company wants to develop a product using new technology in which they have no experience. Whether they would be successful in developing and launching this product is highly uncertain. If they were to fail, they would lose the $100 million invested. They can reduce risk and improve their chances of success if they acquire a firm that already has experience with this market and technology, instead of developing the product in-house. Alternatively, they can pass on the product and eliminate all risk. They might evaluate their options as follows:

- **High risk.** Adopt in-house development, where we lack the necessary expertise

- **Medium risk.** Acquire the required capabilities

- **Eliminate risk.** Do not pursue the project

Note: Often, the degree of risk is subjectively rated on a scale of 1 to 5, with 1 representing extreme risk (high uncertainty and/or high amount of risk) and 5 representing no risk (no uncertainty and minimal risk).

Risk Attitude

The company's risk attitude is its willingness to take on risks in order to obtain profits.

Example: Virgin Galactic, aiming to provide commercial flights in space, is risk-seeking. They are investing heavily in developing new products and technologies. Whether they will be successful is highly uncertain, but they are betting they will succeed and earn a substantial return on their investment.

Note: Often, risk attitude is a subjective description ranging from "risk-averse" to "risk-seeking." One way of describing risk attitude is to determine how much (time, dollars, people) the organization is willing to invest at different risk levels.

Unk-unks (unknown unknowns)

Sometimes, you don't know what you don't know. These are unidentifiable risks that you know nothing about and don't see coming. You can only identify them after the fact. They are most likely to occur in the part of the project you know the least about and where system interfaces occur.[26]

Example: A product in development requires a unique component only produced by a single supplier. In a surprise announcement, the team learns a competitor is acquiring that supplier, which may mean that the component is no longer available in the future.

Notes

1. AJ Justo, "The Knowns and Unknowns Framework for Design Thinking," *Medium*, UX Collective, February 17, 2019, https://uxdesign.cc/the-knowns-and-unknowns-framework-for-designthinking-6537787de2c5; Slavoj Zizek, "Rumsfeld and the Bees," *The Guardian*, June 27, 2008, https://www.theguardian.com/commentisfree/2008/jun/28/wildlife.conservation.

2. U.S. Chamber of Commerce Foundation, *Enterprising States 2015: States Innovate* (Washington, DC, 2015).

3. David Smith and Craig Mindrum, "How to Capture the Essence of Innovation," *Accenture Outlook Journal* 1 (January 2008): 1–10.

4. Sylvain Lenfle and Christoph Loch, "Lost Roots: How Project Management Came to Emphasize Control over Flexibility and Novelty," *California Management Review 53*, no.1 (Fall 2010).

5. Stage-Gate International, "Stage-Gate® – Driving Performance for 30+ Years," infographic, accessed April 17, 2017.

6. Stage-Gate International claims that, by about 2007, 73 percent of North American companies had adopted Stage-Gate, and by about 2014, 80 percent of Global 1000 companies had adopted Stage-Gate. See infographic. Stage-Gate®—Driving Performance for 30+ Years." See also R. G. Cooper and S. J. Edgett, "Best Practices in the Idea-to-Launch Process and Its Governance," *Research-Technology Management* 55(2) (2012): 43–54.

7. Robert G. Cooper, "The Stage-Gate Idea-to-Launch System," in *Winning at New Products: Creating Value through Innovation* (New York: Basic Books, 2011).

8. Joe Tidd and John Bessant. *Managing Innovation: Integrating Technological, Market, and Organizational Change*, 2nd ed. (Chichester, UK: John Wiley & Sons, 2001).

9. Cooper, "The Stage-Gate Idea-to-Launch System."

10. Rita Gunther McGrath and Ian C. MacMillan, "Discovery-Driven Planning," *Harvard Business Review* (1995): 44–54. Shenhar, Aaron J., and Dov Dvir. "Toward a Typological Theory of Project Management," *Research Policy 25*, no. 4 (1996): 607–32, https://doi.org/10.1016/0048-7333(95)00877-2; Steven C. Wheelwright and Kim B. Clark, *Revolutionizing Product Development, Quantum Leaps in Speed, Efficiency, and Quality.* (New York: Free Press, 1992).

11. Ashok Gupta and David Wilemon, "Changing Patterns in Industrial R&D Management," *Journal of Product Innovation Management* (November 1966), https://doi.org/10.1111/1540-5885.1360497.

12. Patricia J. Holahan, Zhen Z. Sullivan, and Stephen K. Markham, "Product Development as Core Competence: How Formal Product Development Practices Differ from Radical, More Innovative, and Incremental Product Innovations," *Journal of Product Innovation Management* 21(2) (2014): 329-345.

13. Richard Leifer, *Radical Innovation: How Mature Firms Can Outsmart Upstarts* (Boston: Harvard Business School Press, 2000).

14. Donald G. Reinertsen, *The Principles of Product Development Flow* (Redondo Beach, CA: Celeritas, 2009).

15. Ibid.

16. Robert G. Cooper, "Next-Generation Stage-Gate: How Companies Have Evolved and Accelerated the System,"in *Winning at New Products: Creating Value through Innovation* (New York: Basic Books, 2011).

17. Stephen K. Markham and Hyunjung Lee, "Product Development and Management Association's 2012 Comparative Performance Assessment Study," *Journal of Product Innovation Management* 30(3) (2013): 408–429.

18. Barry Jaruzelski, Volker Staack, and Brad Goehle, "Proven Paths to Innovation Success," *Strategy + Business*, Winter 2014.

19. Preston G. Smith, *Flexible Product Development* (San Francisco: Jossey-Bass, 2007).

20. Steve Blank, *Technology, Innovation, and Modern War – Introduction*, September 10, 2020. https://steveblank.com/category/corporate-govt-innovation/

21. Mary Drotar, Only 15-20% of employees actually get it: Memorable moments at the Lean Startup conference, *Strategy 2 Market*, December 18, 2014. https://www.strategy2market.com/items/only-15-20-of-employees-actually-get-it/

22. Katherine Radeka and Tricia Sutton, "What Is 'Lean' about Product Development? An Overview of Lean Product Development," *Visions 31*, no. 2 (2007).

23. Lenfle, "Lost Roots: How Project Management Came to Emphasize Control over Flexibility and Novelty."

24. Ibid.

25. Nassim Nicholas Taleb, The Black Swan: *The Impact of the Highly Improbable*, 2nd ed. (New York: Random House, 2010).

26. Smith, *Flexible Product Development*.

Chapter 2

High-Level Overview of ExPD

Chapter 2 Contents

What to Expect

This chapter provides a high-level overview of Exploratory PD® (ExPD), including a summary of the product development systems approach. Within the ExPD process, we provide an overview of its three major segments: (1) Strategy, (2) Ideas & Selection, and (3) Explore & Create.

1. The Strategy segment provides a guide for the organization to follow by defining how product categories, technological capabilities, and other assets will evolve over time.

2. The Ideas & Selection segment serves to generate, collect, and select the best product ideas.

3. In the Explore & Create segment, a cross-functional team identifies, evaluates, and prioritizes the most impactful uncertainties. The project team then resolves these uncertainties while delivering a market-ready product.

We close this chapter with a discussion of ExPD's most essential component: people.

1 Strategy

2 Ideas & Selection

3 Explore & Create

A Systems Approach

Product development is a complex endeavor, touching almost every department in an organization. It requires a system that integrates the management and coordination of strategy, portfolio management, metrics, market understanding, people, and process (Figure 2.1).

Product development is complex and needs to be managed as a system.

Figure 2.1: ExPD Product Development System

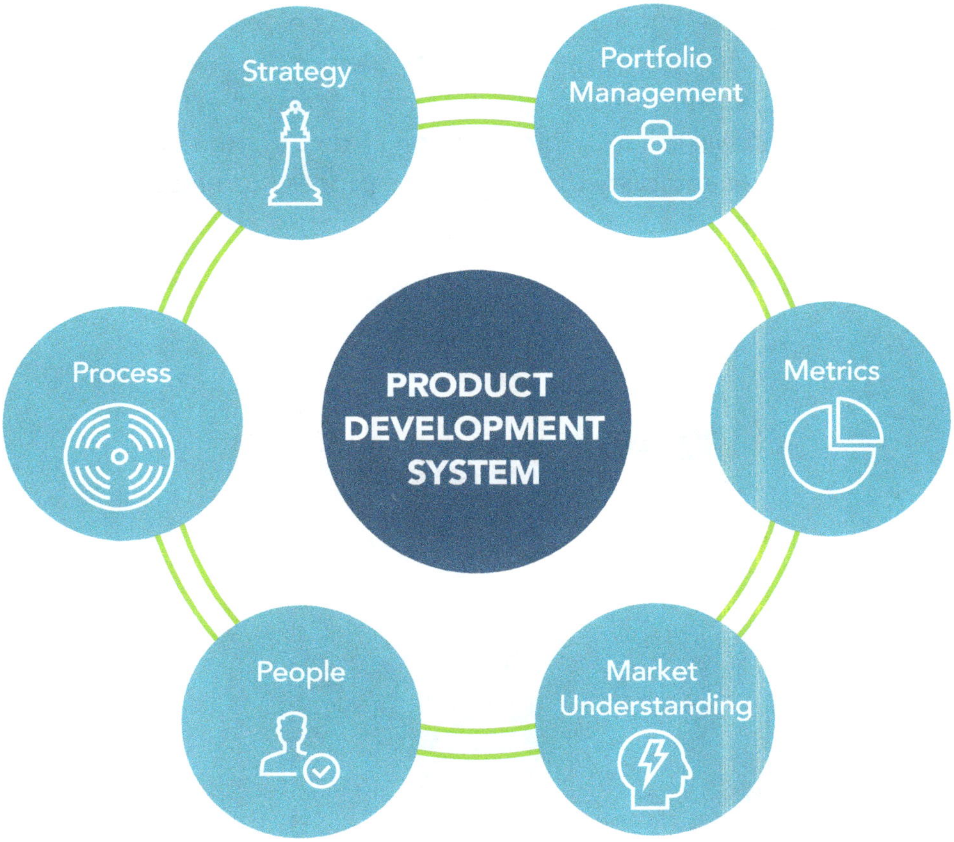

Managers are typically pressed for time and resources and tend to limit their focus to the product development process, with some or all of the following consequences:

- Absence of a strategy to guide product development

- Lack of a defined and strategically integrated portfolio and resource management system, resulting in misguided project selection, an unbalanced portfolio, and inadequate project staffing

- Insufficient infrastructure, including the lack of tools and metrics to support product development

- Inadequate understanding of the market to identify important customer needs

- Misalignment of goals within the organization and team, including a lack of cross-functional integration that leads to conflict and mistrust

- A slow, inflexible product development process leading to missed launch dates, missed sales goals, internal conflict, budget overruns, and release of the wrong products to the marketplace

All the elements of the system must support and reinforce each other to generate new products as effectively and efficiently as possible. The elements are essentially the same for every organization. Still, how the elements are defined, link together, and the specific tools, processes, and skills utilized will differ by organization. These system elements provide the foundation and enable the adaptability of ExPD so the process can respond quickly to changes in internal and external forces.

This is a macro view of ExPD (Figure 2.2). It consists of three segments, Strategy, Ideas & Selection, and Explore & Create. The triangle at the bottom of the process depicts the progressive decrease in risk as the product matures.

We can broadly describe the purposes of each segment as follows:

1. **Strategy.** Establishes a guide for making decisions about which products, markets, and technologies should be pursued.

2. **Ideas & Selection.** Generates, collects and selects the best product ideas.

3. **Explore & Create.** Creates a successful market-ready product while reducing the risk of failure.

In the sections that follow, we take a closer look at each of the three segments.

Figure 2.2: Three Segments of ExPD

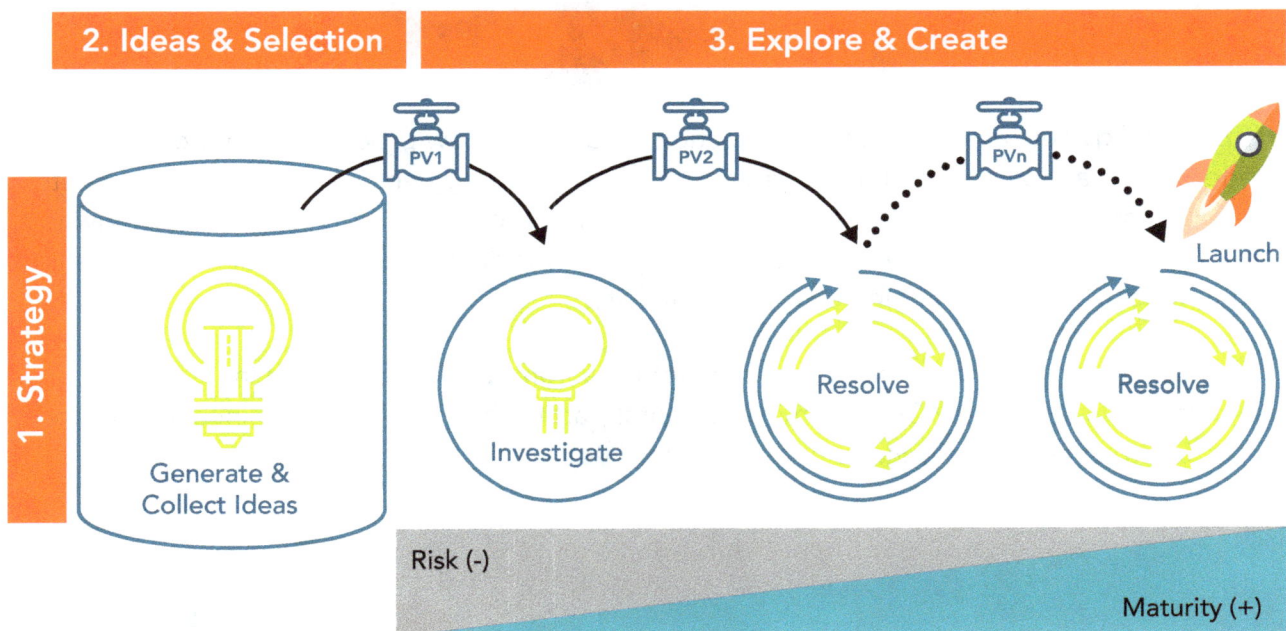

Segment 1: Strategy Segment

The Strategy segment is tightly integrated into the ExPD process and fulfills the need for a common, coherent understanding of goals and objectives. In the 2012 Comparative Performance Assessment Study, 76 percent of the "best" organizations employed strategies to direct and integrate their product development programs.[1] This is not surprising since strategy is positively associated with product innovation for several reasons.[2]

Strategy is positively associated with product innovation.

The strategy sets an organization's direction by defining how product categories, technological capabilities, and other assets evolve. The strategy ensures that the product fits the enterprise's goals and capabilities. If product teams know their organization's strategy and goals, they will spend less time second-guessing what products they should be focusing on, leading to less confusion and increased productivity.[3]

To facilitate strategy integration, we developed a framework called the s2m Strategic Framework™ (see Figure 2.3). It is an instrumental part of the ExPD process since it facilitates adaptation to change and uncertainty.

Figure 2.3: s2m Strategic Framework

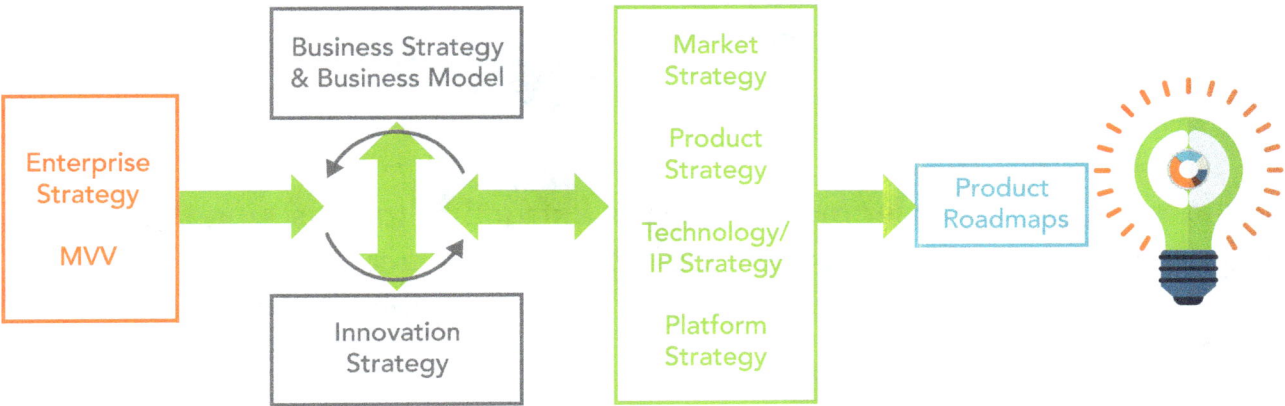

External environments in which an enterprise operates are generally beyond its control. Therefore, the only way to cope is to change strategy to allow the enterprise to survive and thrive. Changing strategy means changing technologies, markets, capabilities, processes, and organizational structures to support the new direction.

In recognizing that changes in the external environment often drive strategy adjustments, we developed guidelines and strategic frameworks for an enterprise to follow based on whether the external environment is Stable, Moderately Dynamic, or Highly Dynamic. While the s2m Strategic Framework differs for each environment, its components remain the same.

For further information on this segment, please refer to Chapter 3, "Why Strategy Matters for Product Development."

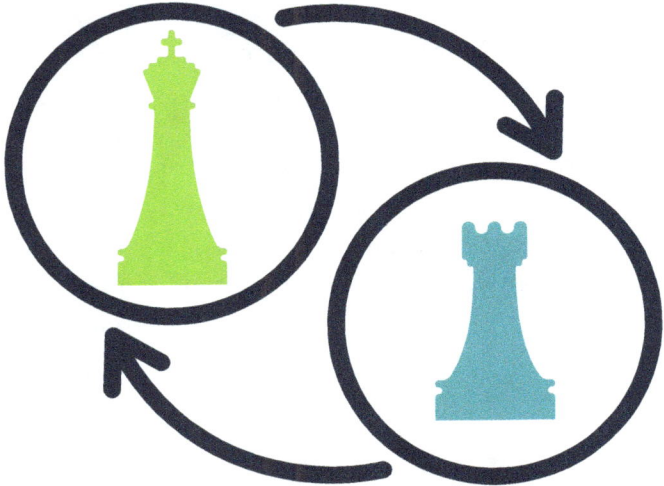

Segment 2: Ideas & Selection Segment

The Ideas & Selection segment provides a structure for ensuring the organization maintains a steady supply of new product ideas. Ideas continuously and systematically generated from various sources provide fuel for organic growth.

Most organizations think they do a respectable job of handling new product ideas, but we often find a lack of any formal, structured process for idea management. In a well-managed product development process, ideas are generated systematically, with well-designed frames, filters, and end-user participation; they are scored, selected, and prioritized consistently; and they are collected and made accessible through an idea library or repository.

Failure to establish an idea management process results in various unproductive outcomes. Companies with an informal idea management approach typically lose sight of ideas that fit strategic needs and opportunities.

Also, without an idea management process, employees may squirrel away product ideas on their hard drives. Consequently, the organization can lose valuable insights and potential intellectual property.

Another symptom is unintentionally revisiting the same product idea repeatedly since the idea's status was not appropriately recorded in the first place.

Failure to establish an idea management process can lead to unproductive outcomes.

The Ideas & Selection segment has four key components (Figure 2.4). The team starts by (A) generating ideas, which it then (B) captures and filters, and then (C) prepares for management review. The fourth component (D) is the unique ExPD feature referred to as Prioritization Valve 1 (PV1). A valve is used to illustrate increasing or decreasing the rate of flow of ideas, based on the available resources. Without this valve, the development pipeline is easily overloaded, and innovation can slow to a crawl.

Figure 2.4: Ideas and Selection Segment

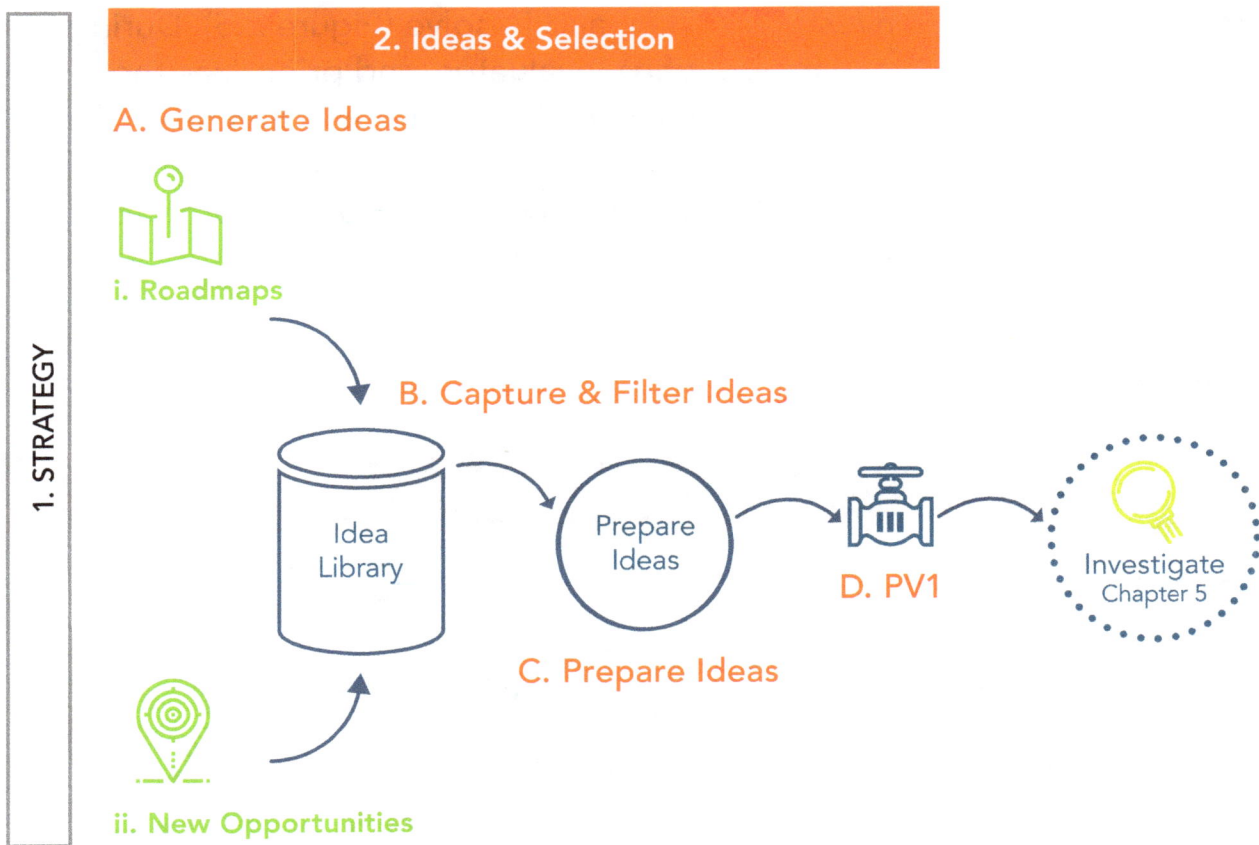

Further information on this segment is provided in Chapter 4, "Idea Management System."

Segment 3: Explore & Create Segment

In contrast to the Ideas & Selection segment, the Explore & Create segment focuses on developing only one opportunity or product idea. It involves three major activities: Investigate, Plan, and Resolve (Figure 2.5). During **Investigate,** the project team identifies, evaluates, and prioritizes the uncertainties and risks that can adversely affect a project. **Plan** helps the team determine the appropriate activities and resources to be executed. During **Resolve,** the project team reduces the most impactful product uncertainties and works towards a launched product.

Figure 2.5: Explore & Create Segment

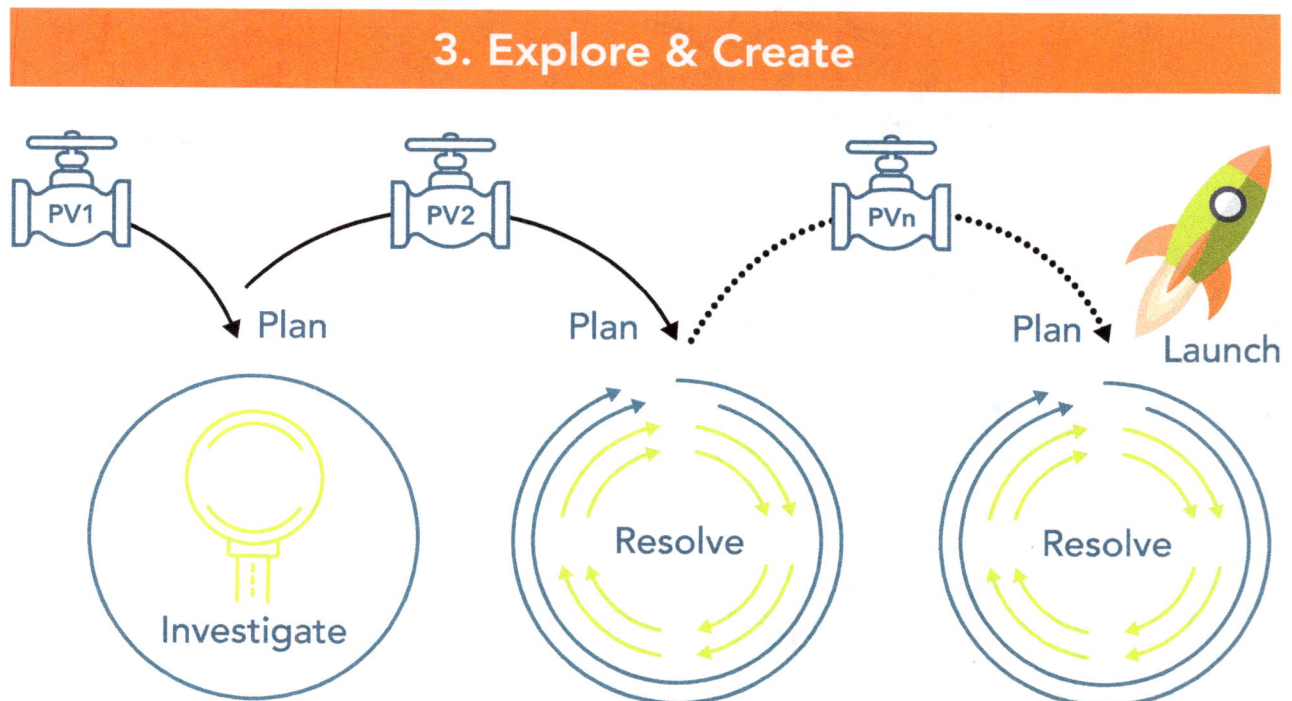

3. Explore & Create

PV1   Plan   Investigate

PV2   Plan   Resolve

PVn   Plan   Launch   Resolve

Investigate

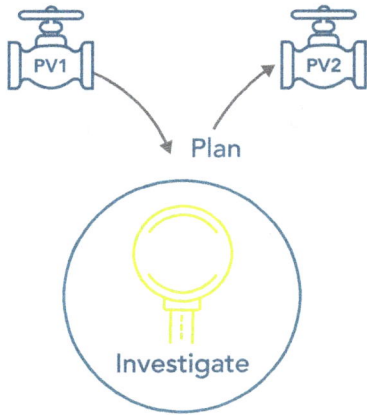

Beginning with the chosen product idea, Investigate splits into two primary tracks of work. In Track A, the cross-functional team identifies, evaluates, and prioritizes the most impactful product uncertainties.

In Track B, the Idea Maturity Model™ is used to identify which development activities are necessary to achieve a launched product. These two major work tracks inform the Product Charter and the Resolve Development Plan (RDP), two essential ExPD planning documents (Figure 2.6).

Figure 2.6: Investigate Process

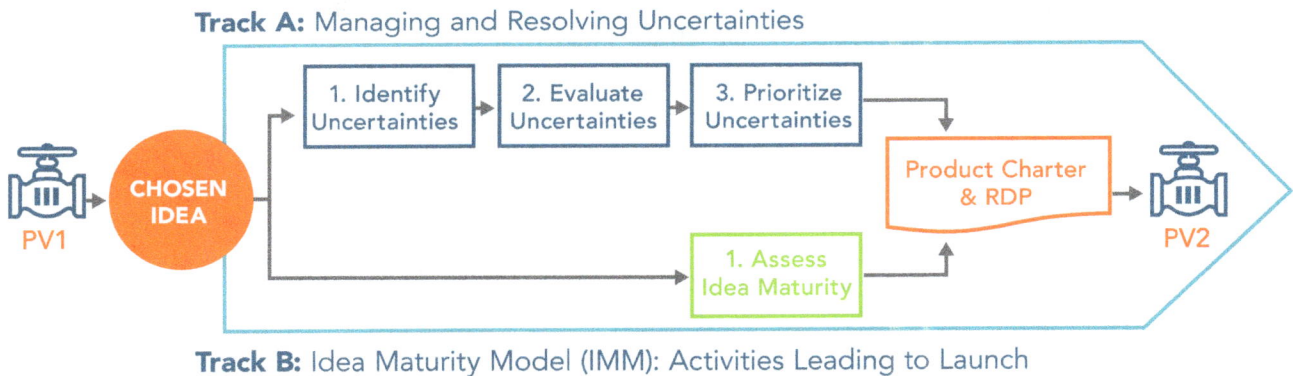

Track A: Managing and Resolving Uncertainties

PV1 — CHOSEN IDEA — 1. Identify Uncertainties — 2. Evaluate Uncertainties — 3. Prioritize Uncertainties — Product Charter & RDP — PV2

1. Assess Idea Maturity

Track B: Idea Maturity Model (IMM): Activities Leading to Launch

Track A: Identify, Evaluate & Prioritize Uncertainties

A cross-functional team identifies, evaluates, and prioritizes the most impactful product uncertainties during Investigate. The team is cross-functional because product uncertainties run across multiple areas within an organization. Its members can be drawn from design, engineering, marketing, product management, operations, supply chain, sales, regulatory and legal affairs—whichever functions make sense within your organization.

Most organizations recognize technology risks only after the product has entered the development phase, through a Failure Mode Effects Analysis (FMEA). In contrast, ExPD addresses risks and their underlying uncertainty very early in the process. Areas of uncertainty can also be adapted for your organization; for example, medical device companies may want to understand the uncertainties relating to regulation and quality.

ID, Evaluate, Prioritize Uncertainties

In Track A of Investigate, when referring to product uncertainties and risks, we recommend using the term "assumptions." While helping to facilitate comparisons between different uncertainties, risks, constraints, and unknowns, the assumption also sets-up the team to prove or disprove a hypothesis.

Refer to uncertainties and risks as "assumptions" for a common framework and positive language.

Track B: Idea Maturity Model

In parallel with Track A, the cross-functional team should follow Track B, which addresses the maturity of the product idea, using the Idea Maturity Model (IMM). The team must understand the project's starting and ending points. We created the IMM so the project team can identify the maturity of the product idea and the associated activities needed to move the idea through the ExPD process until it is launched successfully in the marketplace. The IMM helps the team gauge the time and resources needed to execute the project at a high level.

Assess Idea Maturity Model (IMM)

It also provides the project team with the necessary rigor and guidance throughout the process to ensure that all the critical development activities are being performed, thereby keeping the team on track.

Additional details are provided in Chapter 5, "Investigate."

Plan

Upon completion of Investigate, the cross-functional team has determined the maturity of the product idea and prioritized the most impactful assumptions. If the project continues in the process, the team starts planning.

Planning facilitates the flexibility, adaptability, and speed needed in product development. In ExPD, "Plan" does not mean creating something fixed. Planning is sufficiently flexible to adapt to changes in the market or technology and frequently occurs throughout the entire process.

Two key documents support the planning process: Product Charter and Resolve Development Plan (RDP). A project team is assigned to create the Product Charter and the RDP for the management committee to evaluate (Figure 2.7). These two documents support the two-tier, iterative planning approach that facilitates ExPD's speed and adaptability.

The Product Charter contains the long-term vision and goals for the final product. The Resolve Development Plan (RDP) is focused on the short-term: the prioritized uncertainties that must be resolved and the activities required to advance the product idea's maturity.

Figure 2.7: Plan Activities

Track A: Managing and Resolving Uncertainties

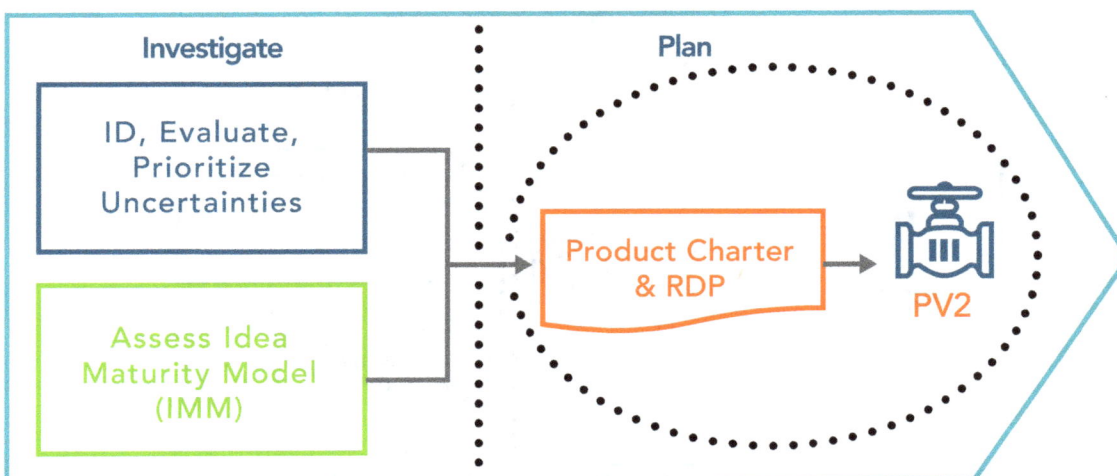

Track B: Idea Maturity Model (IMM): Activities Leading to Launch

At PV2, the management committee obtains additional information from the Product Charter and RDP to support their decision-making. For instance, the project team may have uncovered additional costs or risks while preparing these two documents. With this additional information, the management committee determines whether to proceed to Resolve. The management committee also determines if sufficient resources are available, and if not, the project is put in a queue until they become so.

As for PV1, PV2 serves to prevent overburdening the system with too many projects. Although both prioritization valves have the same purpose, the extent of resources committed at PV2 is greater. PV2 is the gateway to the Resolve Loop.

Additional information on this topic is provided in Chapter 6, "Plan."

Resolve Loop

Once the project clears PV2, the project continues into the Resolve Loop. Inputs to the Resolve Loop include the Product Charter and RDP. The two tracks of resolving and developing continue to run simultaneously throughout the ExPD process, as described in Figure 2.8.

Figure 2.8: Resolve Activities

Track A: Managing and Resolving Uncertainties

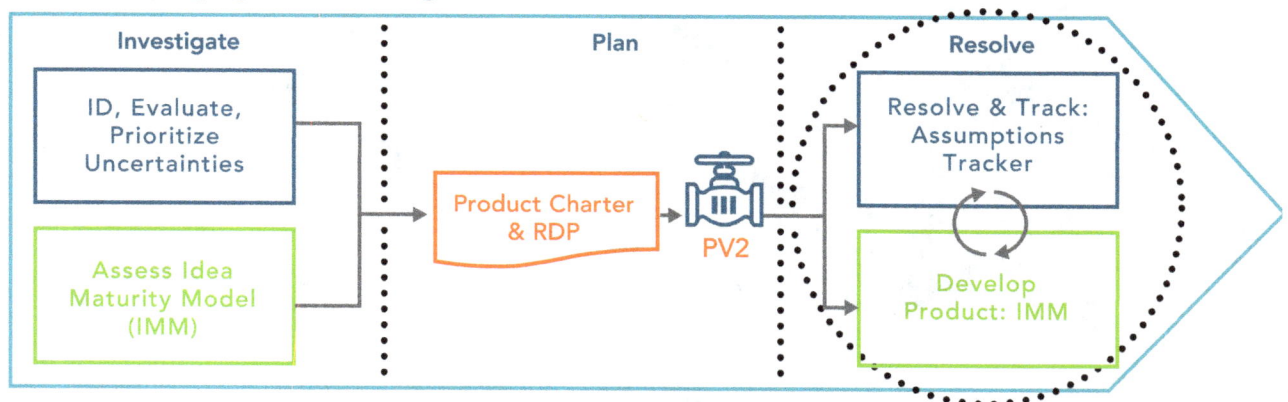

Track B: Idea Maturity Model (IMM): Activities Leading to Launch

Track A: Managing and Resolving Uncertainties

Track A contains a series of Resolve Loops (Figure 2.9). Each addresses an assumption that needs to be resolved by the project team. Generally, the riskier the project, the greater the number of assumptions to be addressed. Resolution can also include testing multiple alternative assumptions in parallel. The activities, the people, and resources required to execute the Resolve Loop depend on what needs to be learned, the resources available, and the budget.

Figure 2.9: Resolve Loops

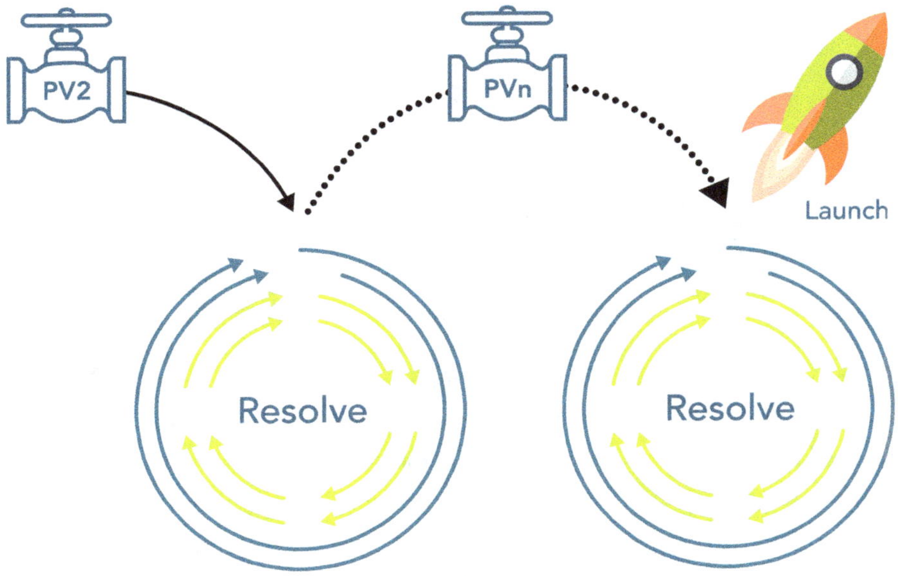

A Resolve Loop, as shown in Figure 2.10, contains four steps, which we call Design, Build, Execute, and Learn & Adapt:

Figure 2.10: Four Steps of the Resolve Loop

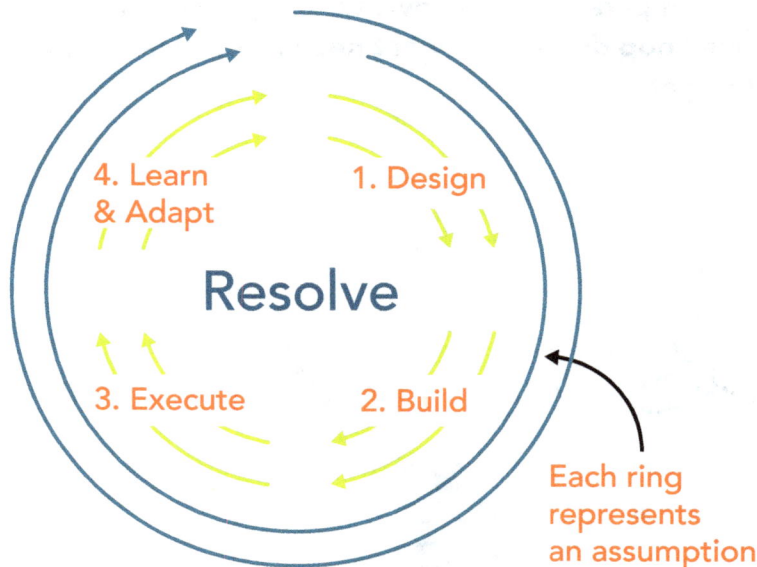

Resolve

4. Learn & Adapt

1. Design

3. Execute

2. Build

Each ring represents an assumption

1. In **Design**, the team finalizes the plan to resolve each prioritized assumption. This step includes but is not limited to constructing hypotheses, reviewing existing data, and identifying the appropriate test methods.

2. In the **Build** step, the team creates the test environment. This may involve testing assumptions by building models, creating prototypes, selecting test methods, or designing surveys.

3. In **Execute**, the test defined in the Design step is performed using the items created in the Build step. Also, the team summarizes and documents their learnings.

4. In the final step, **Learn & Adapt**, the team analyzes the findings. The project team assimilates what has been learned, determines gaps, articulates the most compelling project issues, determines the next steps and adapts as needed.

We like to run Resolve Loops as short iterations, similar to agile techniques. Some assumptions require iteration. While other assumptions will drop off because they're no longer relevant, others will need to be continually monitored, and new ones will be added as they arise. The length of a Resolve Loop varies based on the assumption's complexity and the activities needed for resolution.

Track B: Development Activities: Idea Maturity Model (IMM)

The Resolve Development Plan identified the activities required immediately to advance the product idea within the IMM. These activities are performed in parallel with the resolution activities in Track A.

When the management committee and project team agree that the most significant risks and uncertainties have been reduced to an acceptable level and the development activities are completed per the IMM, a decision is made on whether the project continues in the process. In some instances, the resolution of risks is not acceptable based on the project parameters, and the project may be canceled. The ExPD process is designed to cancel or adapt inappropriate projects before expensive development costs are incurred.

Go to Chapter 7, "Resolve," for more information.

People

The final element of the ExPD approach, People, encompasses four non-discrete areas: Enterprise, Management, Culture, and Teams (Figure 2.11). Distinguishing between these four areas can be difficult. However, the inescapable fact is that it is people who shape your enterprise, who make up your management, who form your culture, and who structure your teams.

People shape your enterprise, make up your management, form your culture, and structure your teams.

Figure 2.11: Interrelationship of People

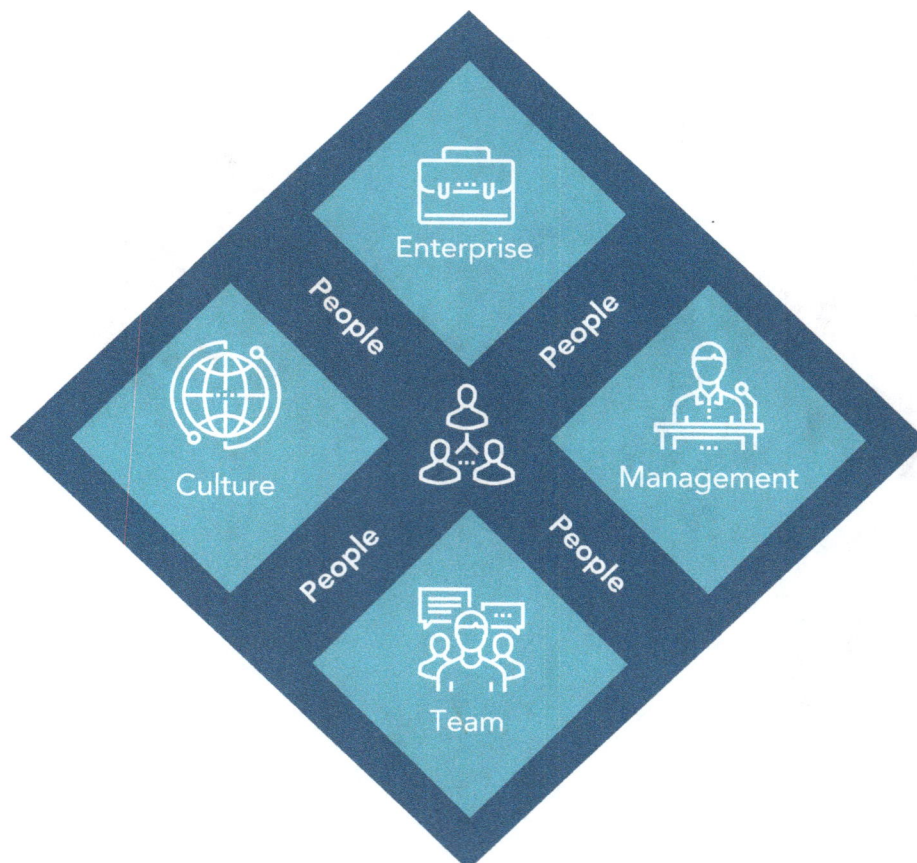

To summarize the significance of these areas, **enterprises** have evolved to a current state of increased complexity, and adaptability is critical for survival. Successful projects have competent **managers** who are involved but not too involved. Some important military techniques, such as mission-type orders, can apply to today's leaders managing product development teams. Other methods that managers can adopt include management by exception and leading by asking questions.

An adaptive **team** structure enables the success of ExPD. This requires striking a balance between structure and flexibility in the workplace. Another issue is identifying what team experience and sophistication are ideal for running an ExPD project team.

For many of our clients, **culture** is a tough nut to crack. It isn't easy to talk about culture without talking about management, teams, and enterprise since they all help shape the culture. Some of the most important aspects of improving culture include adopting better business practices and processes, and addressing the culture of risk-taking, mistakes, and the stigma of failure.

For more information on this topic, go to Chapter 8, "People."

In the next chapter, we will discuss our first segment in the ExPD process, Strategy.

Key Chapter Points

1. Product development needs to integrate all the critical system elements, including strategy, portfolio management, metrics, market understanding, people, and process.

2. ExPD consists of three major segments: Strategy, Ideas & Selection, and Explore & Create.

3. Strategy sets an organization's direction by defining how product categories, technological capabilities, and other assets evolve.

4. The Ideas & Selection segment provides a structure for ensuring the organization maintains a steady supply of new product ideas to fuel organic growth.

5. Two key documents support the planning process: Product Charter and Resolve Development Plan (RDP). The Product Charter contains the long-term vision, while the RDP focuses on the short-term project tasks.

6. The Explore & Create segment involves two major activities, Investigate and Resolve. During Investigate, the project team identifies, evaluates, and prioritizes the product uncertainties and risks. During Resolve, the project team works on resolving the most impactful product uncertainties and develop a launched product.

7. People within an enterprise play an essential part in ensuring the success of product development. People shape your enterprise, make up your management, form your culture, and structure your teams.

Notes

1. Stephen Markham and Hyunjung Lee, "Product Development and Management Association's 2012 Comparative Performance Assessment Study," *Journal of Product Innovation Management 30(3) (2013)*.

2. Michael Song and Yan Chen, "Organizational Attributes, Market Growth, and Product Innovation," *Journal of Product Innovation Management* 31(6) (2014): 1312–29.

3. Chip Heath, "On the Social Psychology of Agency Relationships: Lay Theories of Motivation Overemphasize Extrinsic Incentives," *Organizational Behavior and Human Decision Processes* 78(1) (April 1999): 25–62.

Part II

How to Do It

Segment 1
Strategy

Chapter 3

Why Strategy Matters for Product Development

Chapter 3 Contents

What to Expect

This chapter provides a high-level explanation of why strategy is important for product development. At its most basic, strategy defines where you are going, how you will get there, and how you will know when you have arrived. Strategy is tightly integrated into the Exploratory Product Development (ExPD) process, and it is the first segment within the process.

We will discuss why strategy matters for product development and why it is often slighted or completely avoided. We then present an overview of why the traditional approaches to strategy no longer work and why there is a need for a new adaptive approach that reacts more quickly to continuous change and uncertainty. As an alternative to traditional approaches, we introduce the s2m Strategic Framework™.

We review the three different frameworks (stable, moderately dynamic, and highly dynamic environments) at a general level, covering (a) the characteristics of each environment, (b) a description of each strategic framework, and (c) the internal capabilities and infrastructure needed for each model. We use the term environment instead of market because it is a broader term that can include regulatory, technology, industry trends, and demographic, social, and economic forces. We end the chapter with our case study, describing Turba's strategies using the s2m Strategic Framework. These strategies will help guide Turba throughout the ExPD process.

Why Strategy Matters

Strategy is a critical component of product innovation for several fundamental reasons.[1] It sets the organization's overall direction, influencing governance, culture, human resources, and values. More specifically, strategy shapes an innovation program by defining how product categories, technological capabilities, and other assets evolve. Our model helps select projects and guide them as they progress through the ExPD process to ensure they fit the organization's goals and capabilities.

Strategy is a critical component of product innovation.

Unfortunately, most companies lack a coherent, broadly shared understanding of the organization's goals and objectives, competitive direction, and infrastructure (people, systems and processes, technologies, culture, and metrics). In other words, many organizations have no strategy.

Since strategy is so important, why do we often find that organizations do it poorly or not at all? A strategy is one of those topics most people prefer to avoid, for any of several reasons:

1. Crafting a strategy seems overwhelming, and they don't know where to start.

2. They believe it will take too much time to develop, and it is constantly changing, so why bother?

3. They don't want to dedicate a full-time staff person or team to create a strategy because everybody is busy.

4. They don't want to ask for outside help for fear of a massive bill from a consultant.

5. Some employees don't understand what a strategy is, and they are afraid this inadequacy will be discovered.

6. The output of a traditional strategy development effort tends to be a boring document three inches thick, and no one bothers to act on it.

There is some truth to all these reasons for avoiding strategy. But we'll let you in on a secret: once our clients have a better process, components, and framework to develop strategy, they are eager to discuss it and implement it. Why? They now have the necessary tools, capabilities, clarity, and confidence to discuss and execute the organization's goals and objectives.

Creating a strategic framework is one of the most critical steps to building a better and more adaptive product development system. As consultants working with product development teams, we have seen that it determines whether the state of the organization will be clarity or chaos:

Clarity: Organizations that have well-defined strategies...

- Exude a sense of purpose

- Generate better ideas, with a closer fit to products, categories, competitors, external forces, and customers

- Make decisions more rapidly without extensive debate and politicking because the goals are clear for all members of the executive and project teams

- Demonstrate better alignment of business, innovation, and product strategies, helping to ensure that the most valuable product ideas are chosen

- Coordinate more effectively across multiple product development disciplines and optimize resource allocation across all projects

Chaos: Organizations without strategies...

- Force teams and product managers to develop their product line and category strategies in a vacuum, trying to infer the direction of the organization

- Neglect to involve important functions in the development of innovation and product strategies leading to a lack of consensus

- Create confusion and conflicting paths, causing paralysis or poor decisions (that seem obvious in retrospect)

- Waste time on debates about which projects should be pursued

- Waste resources across the entire organization as the various functions try to cope with different agendas

- Seldom allocate resources optimally— for example, underfunding the most valuable projects and letting valuable projects get preempted as new priorities become urgent

A major reason strategy can be challenging is that so many functional areas (for example, design, engineering, product management, marketing, and operations) are involved in product development. Our findings are supported by a study published in Harvard Business Review that found organizations executing strategy often do well at coordinating up and down in the organization but find coordinating across departments one of the biggest obstacles to successful execution.[2] The authors report, "Only 9% of managers say they can rely on colleagues in other functions and units all the time."

Strategy can be challenging because so many functional areas (for example, design, engineering, product management, marketing, and operations) are involved in product development.

Lack of coordination and communication of product development strategies led us to develop the s2m Strategic Framework. One of the most significant benefits we see is alignment across the various departments by setting clear goals and priorities for product development. This is especially critical in enabling organizations to react and adapt quickly to today's fast-changing markets.

Limitations of Traditional Strategy Development

A strategy is built upon certain assumptions about the future external environment and what the organization can control. For example, the external environment includes competitor activity, customer needs, technological developments, economic trends, and regulatory changes. The enterprise can control what capabilities it maintains, the products and assets it invests in, the markets it serves, and how the organization is structured. Organizations deal with external changes by changing strategy.

Organizations deal with external changes by changing strategy.

Traditional strategy development[3] assumes that markets and industries are fundamentally stable, predictable, and slow to change.[4] Early in the 20th century, this was generally true. More recently, external forces such as globalization and technological advances have enabled rapid, discontinuous, and simultaneous change in most aspects of society, government, and business. Assumptions about the future are uncertain, increasing the risk that strategies will be ineffective or even detrimental.

A traditional approach to creating strategy has four characteristics that do not work well in changing environments:

- Assumes a stable environment

- Focuses internally

- Uses a waterfall approach

- Develops strategy with a few influential experts and stakeholders—the "chosen few"

When organizations **assume the environment is stable,** characterized by a slow and predictable evolution, it makes sense for an organization to look for a position to hold. It methodically seeks that position with a three-step process: (1) analyze the industry and market; (2) determine the best long-term position; and (3) build the organization to own and aggressively defend its position. The organization in a stable environment makes a significant investment in this predetermined position because of the predicted long-term benefit. But in an environment characterized by constant and largely unpredictable change, does it make sense to invest heavily in a predetermined position? Probably not. It's more critical to be able to adapt continuously as the environment changes.

A traditional strategy also **focuses primarily on internal factors,** such as optimizing the organization's investments in operations and maximizing efficiency. Only a few external factors are perceived as critical. An organization may be watching the competition to defend and respond. And it may be paying attention to regulations because that is required. Other external factors are typically ignored, which can be detrimental if an organization is in a changing environment.

The third characteristic, a **waterfall approach,** means the strategy is set at the organization's top levels and cascades down from the enterprise to the business unit to product lines in successively more detailed strategies. This approach is slow and assumes the top levels have the best knowledge about internal and external factors, and it has served companies well during times of slow, predictable change. When the environment is changing, however, the weaknesses of the waterfall approach become apparent. Signs of change are usually seen first by those closest to the market, not senior management. Effective responses are often more apparent on the ground, nearer to customers. A top-down approach to strategy is slow in identifying and reacting to change. Speed and appropriateness of response are critical because the strategic advantage goes to competitors attuned to the market and who can respond rapidly.

A top-down approach to strategy is slow in identifying and reacting to change.

Finally, in a traditional approach, a strategy is often developed by the **"chosen few."** These are usually characterized as MBAs from prestigious business schools or consultants at global firms who work closely with the CEO in an ivory tower on the top floor. This narrow approach to strategy development is detrimental because these individuals are typically removed from the organization's pulse, plus it hampers coordination across the different disciplines. This leads to a lack of integration, differing perspectives, and a lack of buy-in of the strategies.

These four characteristics are the major limitations of traditional strategy. In today's world, many companies are experiencing the impact of these limitations as their businesses are forced to respond to new technologies, new global competitors, the creation of whole new industries, and changing customer needs. The problem has several symptoms:

- Loss of sales and/or market share

- Declining success of new product launches

- Failure to anticipate moves by competitors

- Failure to anticipate the entry of nontraditional competitors

- Disruption from new technologies and business models

Organizations in this situation tend to experience confusion, a pervasive feeling of being directionless, and the common lament, "We need to do things differently." They need a new approach to strategy so they can handle change, uncertainty, and adaptation. We propose the s2m Strategic Framework as an alternative approach (Figure 3.1). It is an instrumental part of the ExPD process and is meant to handle change, uncertainty, and adaptation.

Figure 3.1: s2m Strategic Framework

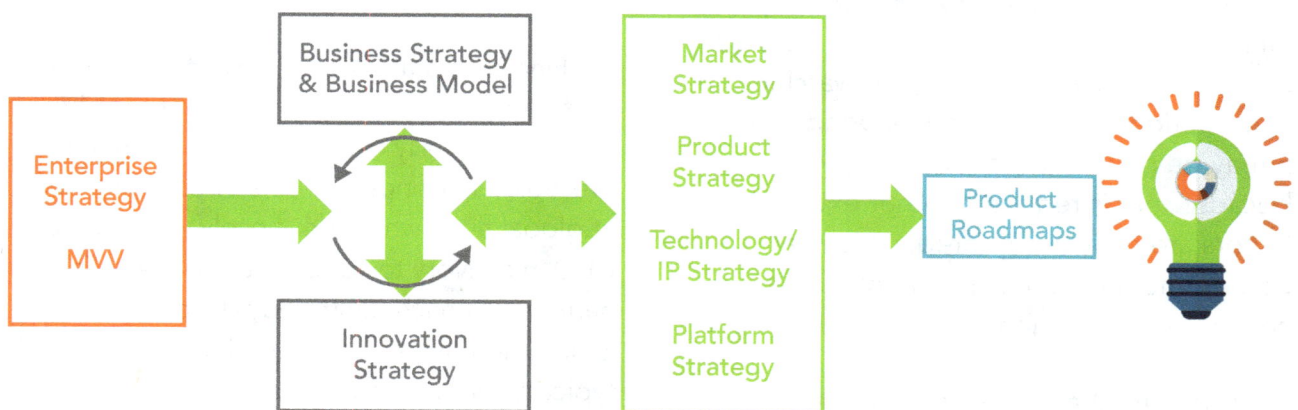

A change in the external environment is frequently the driver to adjust strategy. Therefore, we established different guidelines for an enterprise to follow based on whether the external environment is stable, moderately dynamic, or highly dynamic. The s2m Strategic Framework is different for each environment, but its components are the same, as you will see in the following sections.

The s2m Strategic Framework and ExPD

The external environment in which organizations operate is generally beyond their control. To cope, they must change strategy as needed to allow the enterprise to survive and thrive. Changing strategy means changing technologies, markets, capabilities, processes, and organizational structures to support the new direction, as illustrated by the arrows in Figure 3.2.

Understanding and communicating strategy changes can get complicated without a

Figure 3.2: The Organization Adapts to Changes in the Environment

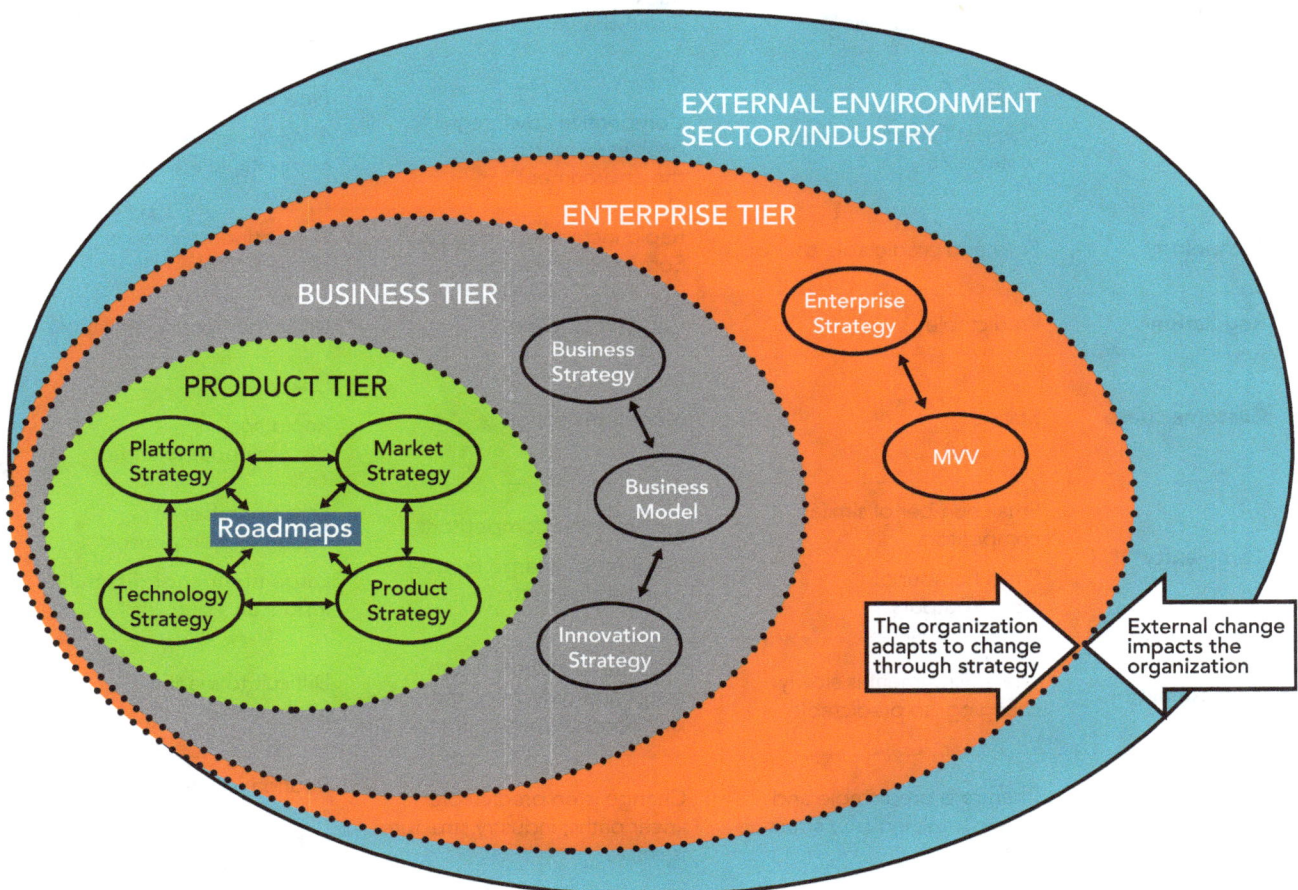

structure or framework to follow. The s2m Strategic Framework organizes the elements of strategy into three tiers: enterprise, business, and product. To create this framework, we classified possible states of the external environment by adapting the classification system identified by Kathleen Eisenhardt and Donald Sull: Stable, Moderately Dynamic, and Highly Dynamic.[5] We applied our three-tier model to describe strategy in each environment. These descriptions are not meant to be definitive, comprehensive, and prescriptive; they are examples used to inform the discussion and development of a strategy. Table 3.1 highlights the significant characteristics of the environments.[6]

Table 3.1: Characteristics of the Three Types of Environments

Environmental Characteristics	Environment Type		
	Stable	**Moderately Dynamic***	**Highly Dynamic**
Demand	Slow and predictable change	Rapid incremental change Somewhat predictable	Discontinuous** change Unpredictable
Competition	Known Competitive advantage is sustainable	Known Competitive advantage evolves to match the environment	New and unexpected Survival versus competitive advantage
Technology	Slow and predictable change	Rapid incremental change Somewhat predictable	Discontinuous change Unpredictable
Regulation	Predictable	Can be complex	Undefined or unpredictable
Customer base	Stable	Moderate change	Unknown Rapid and discontinuous change
Complexity*	Small number of similar competitors Small number of inputs/outputs	Fairly similar competitors Medium to a large number of inputs/outputs	Large number of dissimilar competitors Large number of dissimilar inputs and outputs
Predictability	The environment is slowly changing, so predictable	Possible to identify future states and determine the likelihood of each	Difficult to identify future states
Market boundaries	Change is predictable and incremental; industry structure is stable	Change is on predictable linear paths; industry structure is stable	Boundaries are blurred, successful business models not determined, and roles of market players in flux

Information/ environmental scanning	Focus monitoring on internal operations for efficiency and maintaining market share Information is generally available	Know what to monitor externally: marketplace, competition, acquisitions, technology Information is generally available	Notice what is going on and respond quickly, as opposed to monitoring Information is time-sensitive and imprecise or unavailable
Innovation types	Product revisions and updates Profitability improvement (e.g., cost-cutting and efficiency)	Innovation achieved through acquisition, new markets, new categories	Innovation achieved through experimentation, monitoring, and responding quickly

*Dynamic/dynamism refers to the rate and magnitude of change when compared with industry trends.

**Discontinuous change is a dramatic shift in direction and trend, sometimes referred to as a jolt (e.g., a discontinuous shift in demand for alternative forms of transportation, like Uber/Lyft).

***Complexity is the number and diversity of competitors and dissimilarity of inputs and outputs required by the industry. For example, the auto industry is complex because it involves a host of inputs, vendors, dealers, components, regulations, etc.

We will review the three different strategic frameworks in more detail (Stable, Moderately Dynamic, and Highly Dynamic) starting with the characteristics of each environment. Then, for each, we describe its strategic framework and the internal capabilities and infrastructure needed.

Stable Environment

A stable environment has the characteristics listed in Table 3.2. Change, if any, is predictable and slow. Organizations in this industry have a strong understanding of their markets, products, and technologies, and significant changes on the horizon are not evident.

Table 3.2: Characteristics of a Stable Environment

Environmental Characteristics	Stable
Demand	Slow and predictable change
Competition	Known Competitive advantage is sustainable
Technology	Slow and predictable change
Regulation	Predictable
Customer base	Stable
Complexity	Small number of similar competitors Small number of inputs/outputs
Predictability	The environment is slowly changing, so predictable
Market boundaries	Change is predictable and incremental; industry structure is stable
Environmental scanning	Focus monitoring on internal operations for efficiency and maintaining market share Information is generally available
Innovation types	Product revisions and updates Profitability improvement (e.g., cost-cutting and efficiency)

Products in stable environments tend to have long life cycles. As a result, customer needs are typically well understood, and existing products meet these needs. The technologies used are mature and optimized. The industry is earning acceptable profits, so there is little incentive for incumbents to shake things up. Entry is not attractive to newcomers, so competition also is stable.

> *Products in stable environments tend to have long life cycles. As a result, customer needs are typically well understood, and existing products meet these needs.*

This scenario sounds like nirvana. But you may ask yourself whether any companies today operate in a stable environment. As consultants, we work with companies that act like they're in a stable environment but are not. They may be so internally focused that they do not recognize changes in external factors like new forms of competition or radical technologies. Sometimes they do recognize change but choose to ignore it. Their sales have been slowly eroding over the years, but the business is still profitable, and complacency sets in for the employees.

For example, consider companies developing and manufacturing made-to-order or Original Equipment Manufacture (OEM) products for a long-term client or government project. They receive orders regularly; competition is limited; change is slow; life is predictable. It looks stable, but that's deceptive. The client or government agency can cut the contract(s) at any moment. Radically new solutions may appear and cross over from adjacent niches, fundamentally changing the value equation. Failing to diversify into other markets, customer groups, or product categories can lead to the enterprise's demise.

Strategic Framework in a Stable Environment

As discussed, we should be very careful about defining an enterprise's environment as stable for the purposes of developing a realistic long-term strategy. Nevertheless, there can be value in describing the strategic framework for a stable environment as a starting point. The strategic framework in a stable environment is relatively immobile, with little or no change. It is commonly internally focused, and it most often cascades from the Enterprise Tier (executives within the enterprise) to the Business Tier (executives within the business unit) and then to the Product Tier (typically product management), in the pattern earlier described as the waterfall approach (Figure 3.3).

Figure 3.3: Strategic Framework in a Stable Environment

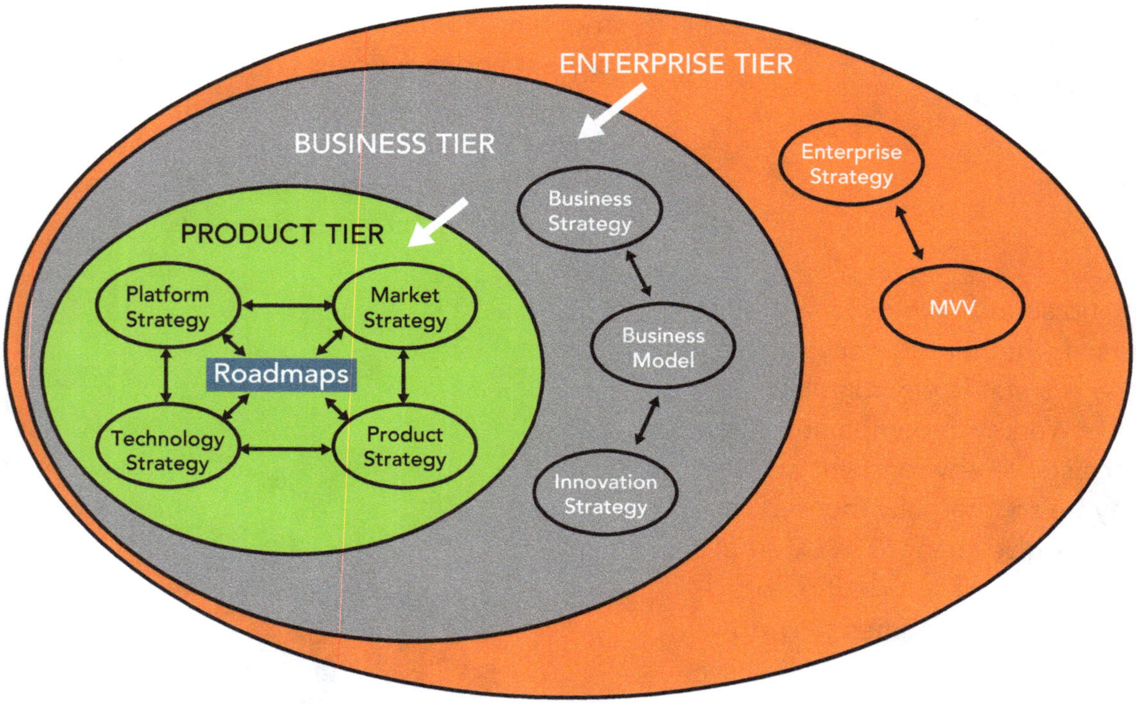

The immobility of the framework is most evident at the Enterprise Tier. There, the enterprise strategy and its Mission, Vision, and Values (MVV) rarely change.

At the Business Tier, the business strategy is developed and typically updated every two to three years. The business model itself stays constant since fundamental change is minimal and incremental change is predictable. Innovation is focused on cost-cutting and increasing market share with product improvements and revisions.

Within the Product Tier, the platform strategy and technology strategy are both focused on driving down cost. Technology is not dramatically different but rather is characterized by simple revisions.

The market strategy focuses on increasing market share and opportunities in adjacent markets. When the environment is stable, competitive advantage can last a long time, so investing in optimizing the competitive position makes sense. Also, when the environment is stable, the organization can define detailed long-term strategies that enhance or reinforce its competitive position.

Internal Capabilities within a Stable Environment

The internal capabilities listed in Table 3.3 provide an overview of the capabilities and infrastructure required for an enterprise in a stable environment.

Table 3.3: Internal Capabilities within a Stable Environment

Traditional strategy	Build and protect a sustainable competitive advantage Stay out of dynamic environments
Source of success	Building and protecting a sustainable competitive advantage Aligning an organization's resources with a competitive advantage
Strategy planning horizon	Long (e.g., 3–5 years)
Product roadmaps	Long (e.g., 3–5 years)
Product strategies	Improving market share/performance
Product development process	Traditional phased-and-gated sequential development
Product development risks	Low risk and uncertainty
Innovation types	Product revisions and updates Profitability through, e.g., cost-cutting and efficiency
Type of learning	Exploitative; honing existing skills and technologies

Environmental scanning	Monitoring of internal operations and competitive position
	Being alert for signs of disruption
Organization structure	Product development under one group
	Exploitative (more extensive and more centralized tight cultures and processes; process optimization is appropriate)
Teams	Siloed
	Traditional lightweight product development teams
Culture	Avoidance of internal competition
	Risk aversion
Metrics	Net present value (NPV); ROI

Strategy in a stable world is about protecting and maintaining what you already have. Since the environment is stagnant, the strategy horizon is typically long, at three to five years. This is also reflected in the product roadmaps, which are typically the same duration.

A traditional phased-and-gated process is suitable for an enterprise in a stable environment if product development focuses on revisions and uncertainty is low.

A traditional phased-and-gated process is suitable for an enterprise in a stable environment if product development focuses on revisions and uncertainty is low.

Cost containment is vital in a stable environment, so lean principles are typically integrated into the process. Learning and skill development focuses on improving and refining current technologies, competencies, and the business model.

The organization structure is traditional, and product development teams typically are siloed or work loosely together as a lightweight team. Heavyweight team oversight is not necessary since products are generally simple revisions or updates. The culture is usually risk-averse and afraid to venture beyond existing market or product categories. To learn more about the differences between lightweight and heavyweight teams, refer to sidebar 3.1.

Sidebar 3.1. The Differences Between Lightweight and Heavyweight Team Structures

This sidebar provides a general overview of the major differences between the lightweight and heavyweight team structures, but there can be variations depending on the organization.

A lightweight team structure typically resides physically within one functional area, and cross-functional coordination is handled by the lightweight project manager through functional area liaisons. Liaisons are typically a middle or junior-level person, and they have little influence in the organization. Lightweight teams are appropriate for simple, low-risk products like a product revision.

Conversely, a heavyweight team structure consists of functional areas typically co-located and assigned to the project full-time. The heavyweight project leader primarily influences the people working on the product development project and can outrank the functional managers. A heavyweight manager is a senior executive with technical expertise.[7] These teams are ideal for high-risk product types.

As mentioned earlier, we see some companies act like they're in a stable environment when they are not. This is probably because, for so long, modus operandi meant traditional strategy development and a stable environment. They have become entrenched in their thinking and their processes. This is when things can go seriously wrong, and companies have difficulty adjusting to the environment's changing needs—think Kodak in the face of digital photography, Blockbuster when Netflix started enrolling subscribers, and taxi cabs when Uber was founded.

Some companies act like they're in a stable environment when they are not.

Given the likelihood of accelerating change and disruption across all industries, companies must prepare for change and adopt some typical practices of a moderately or highly dynamic environment, as described in the following two sections. It's essential to watch for changes and trends that can disrupt your business.

Moderately Dynamic Environment

In a moderately dynamic environment, you find the characteristics listed in Table 3.4. In this environment, the industry is well understood but more complex.

Table 3.4: Characteristics of a Moderately Dynamic Environment

Environmental Characteristics	Moderately Dynamic
Demand	Rapid incremental change Somewhat predictable
Competition	Known Competitive advantage evolves to match the environment
Technology	Rapid incremental change Somewhat predictable
Regulation	Can be complex
Customer base	Moderate change
Complexity	Fairly similar competitors Medium to a large number of inputs/outputs
Predictability	Possible to identify future states and determine the likelihood of each
Market boundaries	Change is on predictable linear paths; industry structure is stable
Information/ environmental scanning	Know what to monitor externally: marketplace, competition, acquisitions, technology Information is generally available
Innovation types	Innovation achieved through acquisition, new markets, new categories

Medical device companies are a good example of organizations that typically operate in a moderately dynamic environment. The industry is highly regulated. Bringing new technologies and products to market is commonly hindered by painstakingly slow and bureaucratic regulatory processes. As a result, change is generally slower and more predictable than in other industries.

Strategic Framework in a Moderately Dynamic Environment

For the strategic framework of companies in a moderately dynamic environment, the strategic tiers (enterprise, business, and product) remain the same as in a stable environment. However, the content of the strategies reflects differences in the environment. The significant difference includes the addition of dynamic capabilities (Figure 3.4). These capabilities enable the organization to adapt to change by altering its business model (see Sidebar 3.1 for a brief description of a business model).

Figure 3.4: Strategic Framework in a Moderately Dynamic Environment

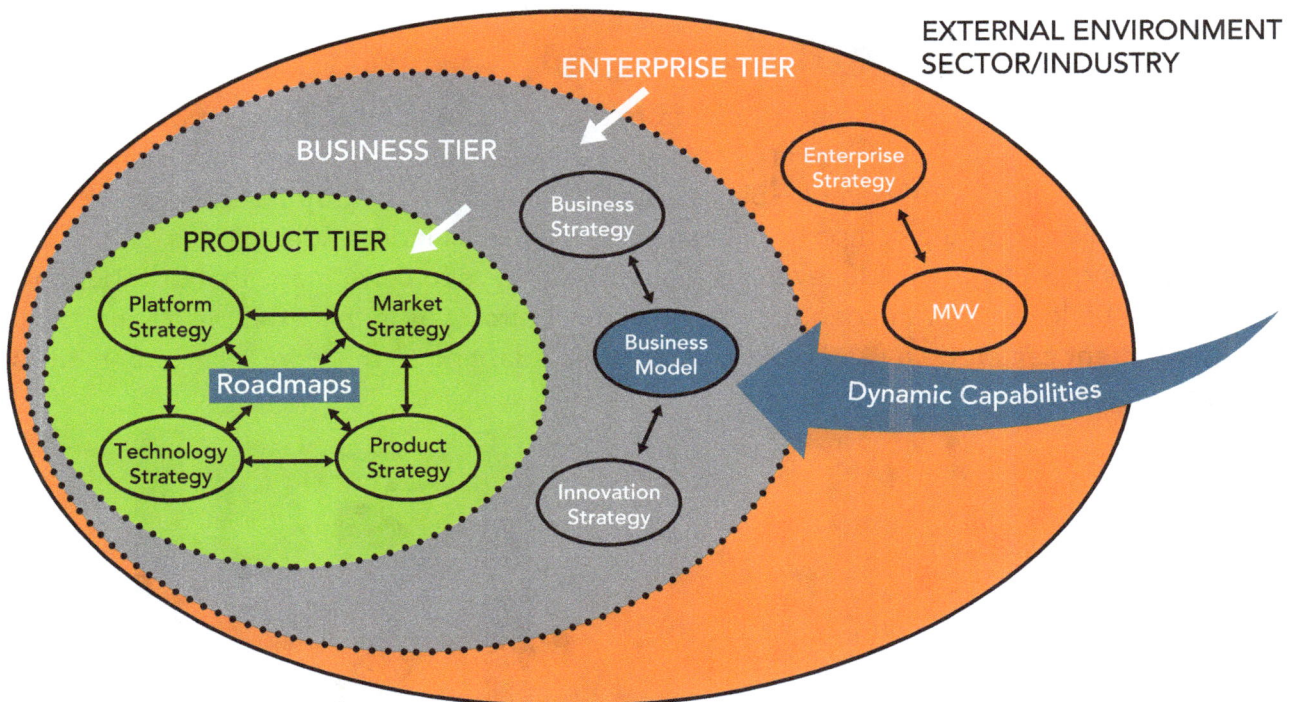

In a moderately dynamic environment, the organization can adapt to change by altering the business model.

Sidebar 3.2. What Is a Business Model?

The business model supports the business strategy by defining the key elements of how the enterprise will create and deliver appropriate products to targeted customers. It identifies the unique choices, processes, assets, activities, technologies, and capabilities that contribute to its products' winning value propositions. The business model makes the business strategy work, and like the business strategy, it should adjust to external market factors.

The business model can support a single product or multiple product lines. With multiple product lines, this implies that the product lines must be similar in important ways, such as similar target markets, similar key technologies, and similar value propositions. Clashes in critical parts of the business model will cause inefficiency at best and value-destroying conflict at worst.[8]

Moderately dynamic environments differ from stable environments. Investing heavily in defending a competitive advantage is generally a bad idea when that investment cannot be turned to a new use. For example, Blockbuster built an extensive distribution network for video rentals, giving the company an advantage over small independent stores. But when streaming became a viable, even preferable alternative to consumers, that distribution network was no longer advantageous. It became an albatross. As in this example, changes in the environment can erode a fixed position's value, so adapting to the external environment is vital, and an adaptive strategy is required.

Internal Capabilities within a Moderately Dynamic Environment

An adaptive strategy builds dynamic capabilities,[9] which enable an enterprise to change. Think of it as changing the organization's business model: the customers it serves, the products it provides, and how it creates and delivers those products.

Product development is a crucial dynamic capability because it can include new markets, products, value propositions, technologies, and skills. In a moderately dynamic environment, the organization can anticipate change and begin building the capabilities it needs (Table 3.5).

Table 3.5: Internal Capabilities within a Moderately Dynamic Environment

Adaptive strategy	Building and maintaining specific processes that enable the enterprise to change its business model and gain a new competitive advantage
Source of success	Competitive advantage is not sustainable in the long-term because the market is changing Adapting the organization to develop new competitive advantages as old ones erode and opportunities for new ones arise Building dynamic capabilities for the kinds of adaptation expected
Strategy planning horizon	Midterm (1–3 years)
Product roadmaps	Midterm (1–3 years)
Product strategies	Ability to identify possible future states, likelihood, and the appropriate types of innovation Development of capabilities to support these types of innovations
Product development process	Combination of traditional process for existing product categories and an adaptive approach, like ExPD for riskier projects
Product development risks	A mix of projects with varying degrees of risk
Innovation types	Innovation achieved through acquisition, new markets, new categories
Type of learning	A mix of exploitative and exploratory (ambidextrous)
Information/ environmental scanning	Monitoring of current markets, products and technologies to identify changes in the environment. A mix of traditional market research and early prototyping with customer input
Organization structure	Existing categories split from new, riskier projects More emphasis on existing categories Separate exploitative and exploratory into different business units

Teams	Cross-functional
	A mix of traditional cross-functional product teams (exploitative) and autonomous teams (exploratory)
Culture	Avoidance of internal competition
	Segregation of risk behavior by unit/teams
	Maintenance of two cultures (exploitative and exploratory), causing friction
Metrics	Dependent on product type

To return to the medical device example, many companies have found building internal expertise in new technologies challenging, hindering the internal development of breakthrough products. Instead, they have developed a new dynamic capability: finding and acquiring companies with desirable technologies and products already in development. This capability enables the acquirer to capture new markets and introduce new categories more effectively.

A challenge of an adaptive strategy is finding a way to manage two fundamentally different activities: (1) maintaining the existing enterprise by improving capabilities and (2) adapting the enterprise by building new dynamic capabilities. The differences involve the activities and learning that take place:[10]

- **Exploitative learning** focuses internally, leveraging the existing product/market knowledge base to incrementally improve current technologies, competencies, and business models. Maintaining the existing enterprise means fine-tuning the existing competitive position and competitive advantage by investing in the current skills and technologies. This is analogous to maintaining the enterprise in a stable environment.

- **Exploratory learning** is more outwardly focused. It uses experimentation to discover new technologies, build new competencies, and create new business models. Adapting the business means learning new skill sets and establishing new processes. These capabilities enable changing the business model as needed.

Exploitative learning focuses internally, while exploratory learning is outwardly focused.

Some innovation experts believe the two types of activities (exploitative and exploratory) cannot coexist in the same business because they compete for the same resources, require different types of employees, and are best supported by different management systems. However, there are a couple of possible approaches to meeting this challenge.[11] One option is for the enterprise to alternate its focus between the exploratory and exploitative innovations in a planned and structured way. An alternative is for the organization to seek ambidexterity and simultaneously pursue both types of innovation by dedicating separate units to exploration and exploitation.[12]

Other experts believe that simultaneously pursuing exploration and exploitation is possible and creates synergy but requires very flexible resources—people who can easily shift between the two types of activities. One approach is managing the enterprise as a "dynamic community,"[13] which has four characteristics:

1. Business units are modular, so they are easily reconfigured.

2. An enterprise culture balances competition and cooperation between business units.

3. Dynamic capabilities are in place for reconfiguring the business units.

4. An organizational structure enables decentralized control but is still directed by enterprise management. The enterprise focuses on spotting new opportunities, reconfiguring the businesses to capture opportunities, and ensuring the appropriate culture. The business units focus on leading their companies as chartered by enterprise management.

It may be necessary for the enterprise to follow two product development processes, depending on the product type. For simple product revisions, a traditional lite phased-and-gated process may be appropriate. For new, riskier products or categories, we recommend ExPD.

Highly Dynamic Environment

A highly dynamic environment has the characteristics listed in Table 3.6. Such markets are characterized by rapid, discontinuous, unpredictable changes, resulting in uncertainty and instability. The organization cannot foresee a future change, and the old ways of doing business no longer work.

Table 3.6: Characteristics of a Highly Dynamic Environment

Environmental Characteristics	Highly Dynamic
Demand	Discontinuous change Unpredictable
Competition	New and unexpected Survival versus competitive advantage
Technology	Discontinuous change Unpredictable
Regulation	Undefined or unpredictable
Customer base	Unknown Rapid and discontinuous change
Complexity	Large number of dissimilar competitors Large number of dissimilar inputs and outputs
Predictability	Difficult to identify future states
Market boundaries	Boundaries are blurred, successful business models not determined, and roles of market players in flux
Environmental scanning	Notice what is going on and respond quickly, as opposed to monitoring Information is time-sensitive and imprecise or unavailable
Innovation types	Innovation achieved through experimentation, monitoring, and responding quickly

Highly dynamic environments are characterized by rapid, discontinuous, unpredictable changes, resulting in uncertainty and instability.

Here are some signs that an environment is highly dynamic:

- Technology is unpredictable, with discontinuous change, meaning a dramatic shift in direction and trend, sometimes referred to as a jolt

- Competition is new, unexpected, and global

- Regulation can be undefined and unpredictable

- Customers can be unknown, new, and changing

- Market boundaries are blurred. Successful business models are not yet determined, and the roles of market players are in flux

- Information is time-sensitive and imprecise or unavailable

One historical example from the 20th century is photography. A stable environment based on film technology was replaced by new technology: digital photography. The new technology enabled a whole new class of competitors and the destruction of unprepared incumbents like Kodak.

More recent examples include new competitors wielding new business models, including autonomous vehicles (self-driving cars). New competitors entered the market from dissimilar industries, such as Waymo, a subsidiary of Alphabet Inc. Lyft and Uber, also entered as solid contenders but pulled out because of the unanticipated complexity of the technology.[14] The auto industry is an important example of how the environment and technology can change quickly and how companies must change and adjust to never-imagined forces.

In a highly dynamic environment, the organization must be flexible and adaptable through quick learning and experimentation as changes in the environment unfold. Simple rules about opportunities worth pursuing replace the detailed foresight and planning you would typically find in a stable environment.[15]

In a highly dynamic environment, the organization must be flexible and adaptable through quick learning and experimentation as changes in the environment unfold.

Strategic Framework in a Highly Dynamic Environment

The s2m Strategic Framework (Figure 3.5) was developed to serve companies in a highly dynamic environment. This strategic framework is quite different from that in a stable or moderately dynamic environment. With traditional strategy, the advantage comes from leveraging resources or hunkering down on a winning market position. In contrast, in a highly dynamic environment, the advantage comes from seizing opportunities quickly.[16]

Figure 3.5: Strategic Framework in a Highly Dynamic Environment

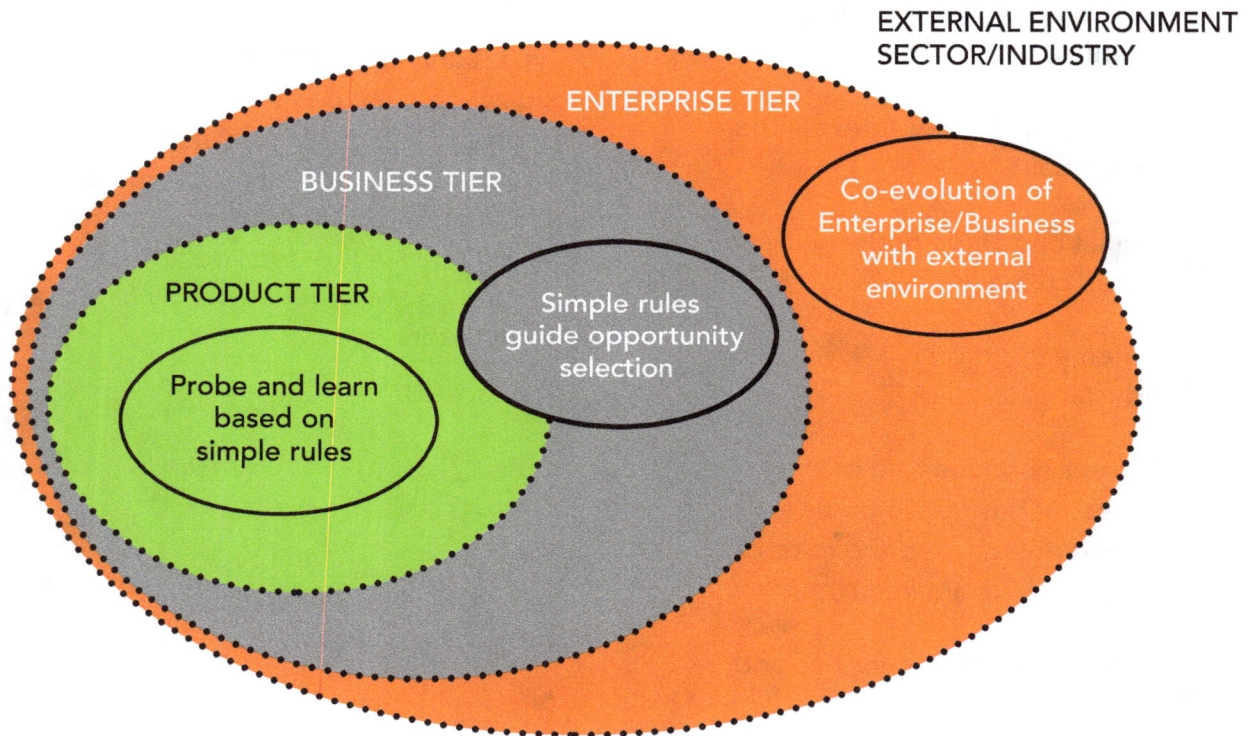

EXTERNAL ENVIRONMENT
SECTOR/INDUSTRY

ENTERPRISE TIER

BUSINESS TIER

Co-evolution of
Enterprise/Business
with external
environment

PRODUCT TIER

Simple rules
guide opportunity
selection

Probe and learn
based on
simple rules

Within this framework are three tiers. The dotted lines in Figure 3.5 indicate permeability between each tier because information and adaptation need to flow between the tiers, including the highly dynamic external environment.

Each tier has its approach to strategy:

1. **Strategy at the Enterprise Tier.** Management determines what businesses to enter or maintain and adjusts that determination as necessary to match changes in the external environment. Adaptability is achieved through coevolution between the enterprise and business units and the external environment.[17]

2. **Strategy at the Business Tier.** The enterprise dictates opportunities to pursue at the business unit level. These opportunities, defined by simple rules,[18] help identify what types of opportunities to pursue, including markets, products, and technologies (Sidebar 3.3). This identification of opportunities helps define the business unit's "playground." Simple rules enable focus and where the business unit should be looking for signs of change.

3. **Strategy at the Product Tier.** Planning the evolution of products and platforms in a highly dynamic environment is challenging since the organization continuously adapts. Therefore, instead of a well-planned product strategy, opportunities are determined by probe-and-learn tactics.[19] Probing helps the enterprise find potential markets with an early version of a product (for example, a storyboard or early-stage prototype) and enables teams to learn from the targeted customer through an iterative process. Experimentation is used to test those opportunities defined by the simple rules at the business unit level.

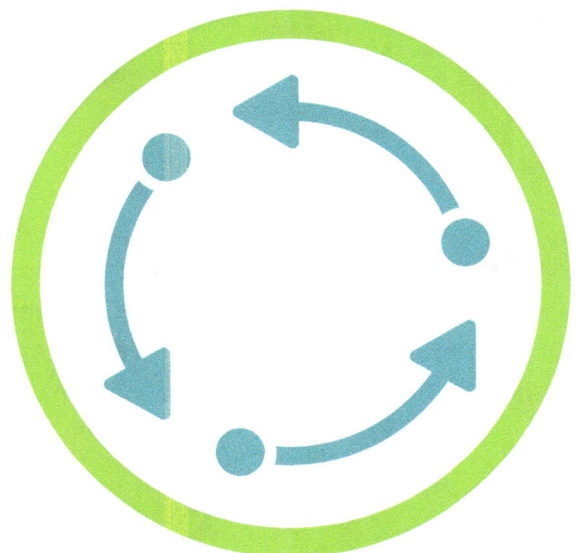

Sidebar 3.3. Simple Rules for Evaluating Strategic Opportunities

In a highly dynamic environment, simple rules help an organization capture unexpected opportunities in the middle of market confusion. This effort starts with the development of a few critical strategic processes. Within these strategic processes are simple rules that give the team direction and parameters for responding quickly to fast-moving opportunities.[20] Table 3.7 provides examples of strategic processes and corresponding simple rules.

Table 3.7. Examples of Strategic Processes and Simple Rules

Strategic Processes	Simple Rules
Growth through acquisition	Acquire companies that have annual revenue of $5–$20 million
	Acquire companies that focus on a particular product category (e.g., orthopedics, robotics, or augmented reality)
	Acquire companies that have the internal resources to support the product for the next 2–3 years
	Acquire companies located in Southeast Asia and Eastern Europe
Excellent customer care	Calls are to be answered on the second ring by a customer service representative (CSR)
	Customers will be assigned to a CSR until their problem is resolved
	Each CSR has a discretionary budget of $200 to appease a customer
40% reduction in time to develop products and 40% increase in product success	Run projects using ExPD
	Run quick-iteration experiments to reduce uncertainty and risk
	Iterate with the customer early and often
	Use prioritization valves to ensure that allocated resources are available before releasing a project into the process

These examples provide simple, clear tactical rules that employees can easily follow, so they can quickly capture short-lived opportunities. The number of rules can range from two to seven; the optimal quantity depends on the organization. Too many rules can restrict employees, while too few can leave employees guessing. When the environment is highly dynamic and less predictable, fewer rules are recommended to increase flexibility.

Internal Capabilities within a Highly Dynamic Environment

Highly dynamic environments call for a continuous-change strategy (Table 3.8). Because the environment changes so rapidly, any competitive advantage can be expected to have a short life. There is limited time for analysis and building traditional forms of competitive advantage. Success comes from seizing the right opportunity and learning as you go.

Highly dynamic environments call for a continuous-change strategy. Because the environment changes so rapidly, any competitive advantage can be expected to have a short life.

Table 3.8: Internal Capabilities within a Highly Dynamic Environment

Continuous change strategy	Pursuing multiple opportunities simultaneously, responding quickly to market opportunities, and learning as you experiment
Source of success	Market changes so quickly and unpredictably that opportunities are short-lived. There is no time for analysis and building traditional forms of competitive advantage. Success comes from grabbing the right opportunities and learning as you go
Strategy planning horizon	Short (1 year or less))
Product roadmaps	Short (1 year or less)
Product strategies	Without the ability to identify future states and likelihoods or to develop technologies and other capabilities in advance, identify, prioritize, and jump on opportunities, guided by simple rules
Product development process	An adaptive approach using exploration and experimentation, like ExPD
Product development risks	High risk and uncertainty
Innovation types	Continuous probing for new opportunities, experimentation, and learning
Type of learning	Exploratory

Environmental scanning	Information is fast paced, limited, quickly outdated and difficult to interpret. Forecasts and historical information are of little use
	Dependence on early prototyping with fast iteration and customer input
Organization structure	Blend of limited structure around responsibilities, priorities, and communication, with the freedom to experiment on projects
	Emphasis on innovation and exploratory units
	Management of multiple projects and transition between projects
Team	Cross-functional
	Autonomous teams with a heavyweight leader
	Fluid job descriptions
	Highly-skilled employees
	One possible organizational model: niched experts working on a contract basis, temporary project teams
Culture	Coexistence of competition and collaboration
	Risk seeking
	Continual change, creating stress
Metrics	Incremental investment (asking, "Is a product worth the next stage of investment?")

A continuous-change strategy requires a supporting infrastructure. Because it is impossible to predict which opportunities will be successful, the organization needs to handle multiple projects while pursuing new opportunities simultaneously. It must be able to experiment and quickly learn about the opportunities. Fast and grounded decisions are paramount, for example, to pursue a new opportunity or cancel an existing project.

Centralized control is too slow in a highly dynamic environment, so we look to the concept of complex adaptive systems proposed by Eisenhardt and Piezunka.[21] A vital feature of this approach is that control is decentralized or distributed, so enterprise and business unit management roles are radically different. The enterprise adapts to changes in the environment through each business unit adapting to the changes in its own market.

Rather than enterprise management being responsible for spotting opportunities and reconfiguring the businesses to exploit them, business unit managers and their teams assume responsibility for spotting and pursuing the best opportunities in their markets.

Business unit managers and their teams assume responsibility for spotting and pursuing the best opportunities in their markets.

Enterprise management creates the culture, structure, and processes that help business units share knowledge, develop relationships, and collaborate. When necessary, they also intervene to optimize the business unit's size and referee competition for the same opportunities.

Business units in a highly dynamic environment are allowed and even invited to compete. The enterprise does not demand cooperation between business units, which can lead to suboptimal performance. Over time, as business units evolve, they may be combined, split, or refocused to improve their ability to spot and leverage opportunities.

Product innovation also is distinctly different. Unlike stable and moderately dynamic environments, where industry knowledge and experience are valuable for product innovation, highly dynamic environments contend with complex, novel, and changing situations. This has several implications for developing products:

- Industry participants must continually monitor and adapt as competitors try out different business models. New competitors from other industries may enter, believing their capabilities are a good fit for this evolving industry

- The situation is also highly complex, novel, and challenging to forecast. Even experienced industry participants can't predict what will be successful

- Exploratory learning through experimentation searches for new combinations of competitive position, competitive advantage, and value propositions that may be successful while keeping in mind that external factors are continually shifting

Complex, novel, and changing situations imply high levels of uncertainty and a high risk of failure. This requires exploratory learning through experimentation.[22]

Complex, novel, and changing situations imply high levels of uncertainty and a high risk of failure. This requires exploratory learning through experimentation.

The product development process must identify the types and sources of uncertainty and risk. The team needs to learn and adapt quickly.

Getting Started: Five Key Tasks

A broad understanding of different competitive environments helps prepare you to get started on establishing a strategy. In establishing a strategy that is appropriate for your organization, it is helpful to think in terms of five key tasks:

1. **Determine what environment your organization is currently operating in.** Table 3.1 highlights the major characteristics across different environments (Stable, Moderately Dynamic, and Highly Dynamic).

2. **Understand the strategic framework recommended for your environment and use this framework to develop your future strategies.**

   a. Strategies require input from multiple disciplines. This enables the team to improve integration, embrace differing perspectives, build buy-in, and improve adaptability and speed.

   b. Product managers and their engineering partners should jointly develop product roadmaps. Refer to the tables identifying internal capabilities, including the appropriate time horizon for your roadmap. The roadmap will feed directly into the next segment of ExPD, Ideas & Selection. The next chapter will provide more information on product roadmaps.

3. **Understand the organizational capabilities required to execute the strategy.** Based on the capabilities in your organization, determine the gaps. Modify and fill these gaps so your organization can react quickly to changes.

4. **Be paranoid.** Get active and involved inside and outside of your industry group. Learn more about dissimilar industries that may disrupt your industry. You can do this by keeping tabs not only on your competition but on adjacent or related industries and the start-up community. Join groups that represent multiple industries and development communities, so you can better understand their capabilities and how they may evolve in the future to affect, challenge, or mirror your organization.

5. **Act fast.** Ensure that your employees understand the necessity of being quick and agile to react to new opportunities.

To see what this approach looks like in practice, read our Turba case study in Sidebar 3.4. Turba's strategies will serve as a foundation for the following case studies.

Sidebar 3.4: Turba's Strategies

TURBA CORPORATION

Turba Corporation is in a moderately dynamic environment and developed its strategies using the s2m Strategic Framework. It began with the Enterprise Tier and then proceeded to the Business and Product Tiers.

Enterprise Tier

Turba's executive team began with the Enterprise Tier, which comprises enterprise strategy and the company's Mission, Vision, and Values (Figure 3.6).

Figure 3.6: Turba's Enterprise Tier

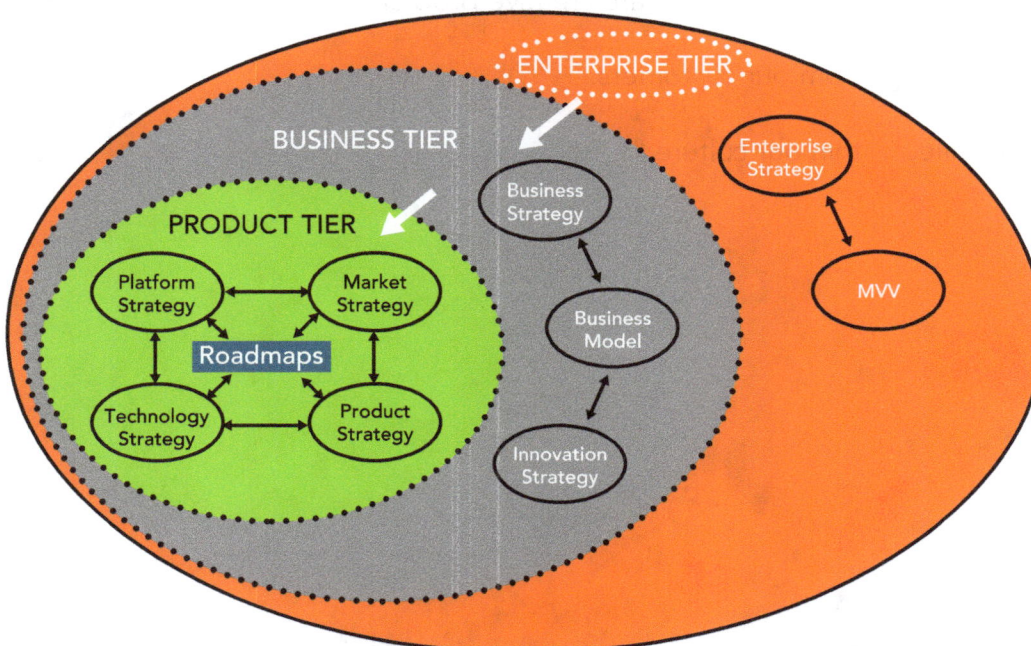

Enterprise Strategy

Turba is a $900 million stand-alone business focused on producing and marketing consumer electronics for home health care. Since it has a single business, it doesn't need an enterprise strategy that addresses multiple business units. Therefore, it will address strategy at the Business Tier. So, the team turns to the company's Mission, Vision, and Values statement.

Enterprise Strategy

Turba's Mission, Vision, and Values

MVV

Turba's mission is to provide convenient, reliable home-based health care through preventive and emergency self-service devices.

Turba's vision is to be the most loved and trusted brand for innovative, reliable, high-quality home health care products.

Turba has identified the following five corporate values:

1. "The highest quality of care for our customers."

2. "Ability to control the quality of the product."

3. "Employees who are adaptable and are creative problem solvers."

4. "Integrate learnings from our customers."

5. "Integrity instilled within our culture."

Business Tier

Turba's executive team then turns to the Business Tier, including the company's business strategy, business model, and innovation strategy (Figure 3.7).

Figure 3.7: Turba's Business Tier

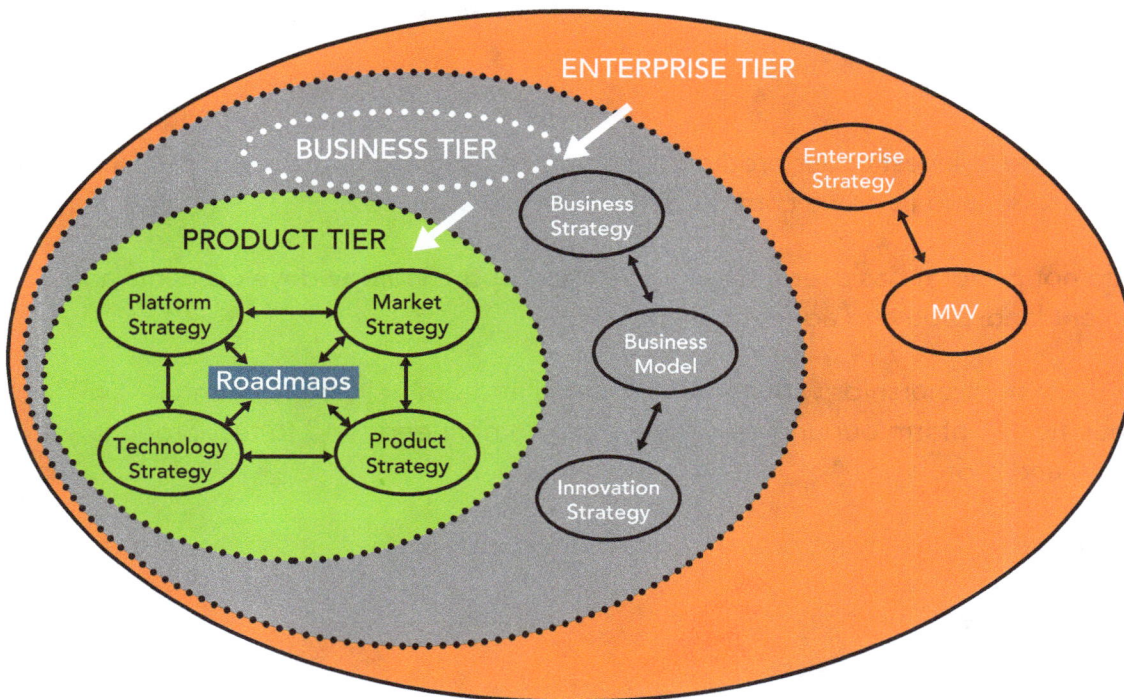

Turba's Business Strategy

Business Strategy

Turba's business strategy is to provide high-quality, premium-price products to the home health care market to increase the user's ability to stay in the home longer (aging in place) and live independently. Specifically, this means Turba will execute their business strategy through the following six tactics:

1. "We will increase our market share by adding new product categories and improving the current products."

2. "We continue to invest in our product development process to reduce uncertainty and time to market, and improve commercial and technical success."

3. "We do not perform basic research, but we monitor technology developments and the patent landscape and adopt as necessary."

4. "We maintain partnerships with home caregivers and senior-living residential communities to inform our understanding of customer needs and to test product concepts and prototypes."

5. "We increase revenue by introducing new product categories."

6. "We hire star talent in critical positions."

Turba's Business Model

Business Model

Generally, a business model is an excellent way to think through the business and understand how it works to create differentiation and constrain the options available to the business. Refer to Sidebar 3.2 for the definition of a business model. Figure 3.8 applies the business model to the PERS that Turba is interested in developing.

Figure 3.8: Turba's PERS Business Model

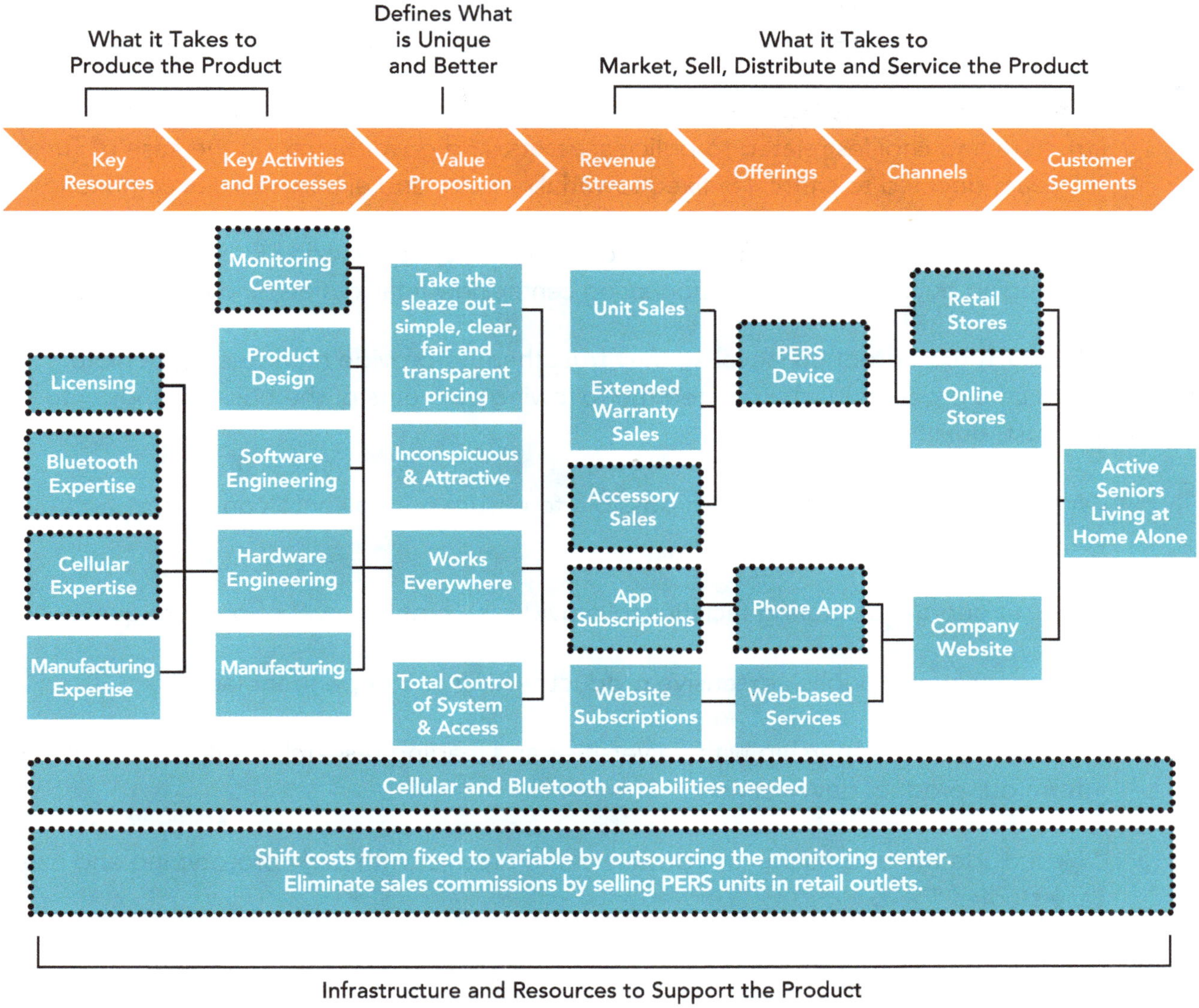

What it Takes to Produce the Product		Defines What is Unique and Better	What it Takes to Market, Sell, Distribute and Service the Product			
Key Resources	Key Activities and Processes	Value Proposition	Revenue Streams	Offerings	Channels	Customer Segments

Monitoring Center

Take the sleaze out – simple, clear, fair and transparent pricing

Unit Sales

Retail Stores

Licensing

Product Design

PERS Device

Online Stores

Bluetooth Expertise

Software Engineering

Inconspicuous & Attractive

Extended Warranty Sales

Active Seniors Living at Home Alone

Cellular Expertise

Hardware Engineering

Works Everywhere

Accessory Sales

Manufacturing Expertise

Manufacturing

Total Control of System & Access

App Subscriptions

Phone App

Company Website

Website Subscriptions

Web-based Services

Cellular and Bluetooth capabilities needed

Shift costs from fixed to variable by outsourcing the monitoring center. Eliminate sales commissions by selling PERS units in retail outlets.

Infrastructure and Resources to Support the Product

The business model highlights the new elements that Turba will need to develop for PERS. It has four major parts:

- On the right hand side, the Revenue Streams, Offerings, Channels, and Customer Segments columns represent what it takes to market, sell, distribute, and service the products

- In the middle is the Value Proposition column. From the customer's point of view, it defines what is different, unique, and better about the business's offerings, allowing the business to gain market share and retain customers

- On the left side, the columns under Key Activities and Processes and Key Resources represent what it takes to produce the product and deliver the value proposition

- At the bottom are the business's infrastructure and resources, including the cost structure and choices related to policies, assets, and governance. In the case of Turba, the team decided to shift from fixed to variable costs and eliminate sales commissions

The dotted lines in Figure 3.8 represent the capabilities needed to add PERS to Turba's business, including but not limited to a monitoring center, new retail outlets, and cellular expertise.

People buy Turba's products because they trust them to provide high-quality, reliable home health care products. Turba created the following elements of the business model to deliver on trust, quality, and reliability:

1. "We choose local suppliers when possible to ensure our ability to control the value chain."

2. "We use our manufacturing facilities to provide the highest level of product quality."

3. "We sponsor and publish extensive product testing at independent labs."

4. "We maintain a partnership with universities and leading research institutes to test and inform our product development."

5. "We include the end-user in the design of the product with early prototyping and use case studies."

Turba's Innovation Strategy

Innovation Strategy

Turba's innovation strategy includes defining the kinds of innovation it will pursue, its approach to innovation, and its infrastructure and investment to support innovation.

Turba's innovation definition: Turba's innovations will provide customers best-in-class reliability and convenience, plus innovative preventive and emergency care capabilities. Ultimately, the goal is to have the user stay in the home longer (aging in place) and live independently.

Turba's innovation approach: Turba is a fast follower that leverages proven technologies and familiar products and markets. They have substantial organic growth goals and realize they need to take on riskier projects to achieve this growth. (The new PERS product category is one of their growth initiatives.) They want to be intentional and careful about the risk they take on, so the ExPD process will be used instead of the current phased-and-gated approach. They will have to make changes to the business model to accommodate the new product categories.

Turba's infrastructure and investment: Turba's innovation infrastructure was appropriate to its fast-follower approach in serving familiar markets and products while adding only proven technologies to its capabilities. The change to entering new product categories, plus adopting a new approach to product development (ExPD) and managing risk, requires modifications to the infrastructure. Table 3.9 lists some areas of change in infrastructure.

Table 3.9: Turba's Innovation Infrastructure: Current and Required

Category of Innovation Infrastructure	Current	Required

People and culture	An organizational structure that encourages cross-functional participation in product development
	
	The hiring of experienced technologists as new technologies are adopted
		Leverage existing cross-functional practices and culture
		
		Hire and retain a mix of star talent and solid performers. Star talent helps identify new opportunities for growth
		
		    Hire Bluetooth and cellular experts
		
		    Hire market intelligence professionals
		
		    Establish part-time Senior Fellow to support ExPD teams
		
		Develop human resources approaches to hiring and retaining star talent
		
		Develop executive skills

Training	On-the-job training focused on the Turba phased-and-gated process for new project team members	Establish new project team and management training related to the ExPD approach: ExPD system, culture, identifying and managing risks, product development strategies, and roadmaps

Capital investment	Test labs for selected technologies
	
	Prototyping labs for selected technologies
	
	Manufacturing equipment and technologies for supported product lines
		Leverage existing test and prototyping labs
		
		    Enhance labs to support ExPD methods of experimentation
		
		    Enhance labs to support new technologies
		
		Leverage existing manufacturing equipment and technologies
		
		Acquire a knowledge management system to capture ExPD assumptions, resolutions, and learnings
		
		Acquire a capacity management/resource optimization tool to support the prioritization valves

Technology acquisitions/licensing	License technologies as needed	Leverage existing licensing process License cellular and Bluetooth technologies

Processes	Phased-and-gated process for product development
	
	Top-down strategic planning
		ExPD for product development
		
		    Requires changes to training, culture, talent, decision-making process, risk management, etc.
		
		Adaptive strategic planning
		
		    Enhance market intelligence capabilities

Product Tier

Finally, the Product Tier involves a strategy for each product line—in this case, the PERS system that Turba is considering. It addresses market, product, technology, and platform strategies (Figure 3.9). For each of these, we present Turba's PERS strategy.

Figure 3.9: Turba's Product Tier

Turba's PERS Market Strategy

Turba's market segment is active seniors living alone at home who want the ability to call for help wherever they are. Seniors living alone and their loved ones share a common fear: The seniors may need help and cannot reach out for help. The nightmare scenario is the senior away from home, hurt, and unable to call for help. There is a need for a PERS that works anywhere, and there is no solution currently on the market.

Additionally, they expect the trend to more health-conscious behaviors to continue. Changing habits can be challenging, and specific reminders to take medicine, drink water, take a walk or go to a doctor's appointment can be very helpful. Because the customer always wears the PERS, it is the ideal device to deliver these critical reminders. It is also important that it has an easy-to-use interface.

They expect the next evolution in health consciousness will be to monitor general health, such as blood pressure, pulse, activity level, and sleep. Health issues are a common fear of active seniors, and they believe health monitoring will help alleviate that fear. They expect this will require an update to the platform.

Goals include:

1. "We will have an overall market share of 20 percent within five years."

2. "Our customer satisfaction rate will be 99 percent or more."

3. "We will achieve revenue of $8 million in the first three years."

Approach:

Turba's approach drives the market by introducing a new value proposition. This unit works anywhere instead of just in the home (commonly delivered through an existing home security alarm/monitoring service). This value proposition is enabled by the innovation of connecting the unit to a smartphone via Bluetooth. The traditional segment has changed, becoming more mobile and tech-savvy, which would help the adoption of the new PERS solution.

This category has experienced rapid adoption with the senior population's increasing size. Turba expects their alternative technology solution, simple user interface, and contemporary design and colors to provide a competitive advantage in attracting newcomers to this category: "Designed for the active senior, who wants an inconspicuous and attractive device." Turba will reinforce its brand positioning of high-quality, innovative products through relationships with leading research institutes.

A new distribution approach will be used to sell units through well-known stores and online retailers. Turba will distance its brand from gimmicky ads.

Infrastructure requirements:

1. "We will execute ongoing market research on competition, external market factors, and market segments."

2. "We will continually gather end-user feedback with early stage prototypes."

3. "We will continuously monitor customer satisfaction and response metrics."

4. "We will establish new sales, marketing, and distribution channels."

Turba's PERS Product Strategy

Product Strategy

Turba will introduce a new line of products in the PERS category. Their goal is to establish a position in the market with a high-end contemporary product with a simple user interface. The product design is modeled on popular wrist health-monitoring devices, but for the senior market.

Approach: Turba will establish a PERS that works anywhere cellular service exists. The initial product introduction will consist of an alarm triggered by the wearer. The alarm is received by a monitoring center (much like a home security alarm) except that an alarm can be sent from any smartphone with Bluetooth capability. It is not tied to the home security system, making it a better fit for wearers who spend time away from home.

Over time, Turba will introduce additional functionality and value through hardware and/or software enhancements specifically tailored to the target segment. Initial extensions will include general wellness maintenance, including medicine and hydration reminders and doctor and lab visit alerts. Additional extensions can consist of health monitoring, such as heart rate, blood pressure, sleep patterns, and cardio fitness.

Infrastructure requirements:

1. "Build capability to remain up to date on the latest design trends and health concerns of our targeted market."

2. "Acquire cellular and related capabilities and tools for product development."

3. "Leverage our manufacturing facilities and supply chain to ensure the highest quality product."

4. "Monitoring center capabilities will be outsourced, so we will need to establish the appropriate metrics to ensure excellent customer care and response times."

The specific elements of the product strategies are illustrated in the multi-level roadmap in Chapter 4 (Figure 4.3).

Turba's PERS Technology Strategy

Technology Strategy

The PERS category will require the acquisition of some new technologies. Their goal is to license the latest technologies which will reduce technology uncertainty. Also, licensing will enable them to enter the market quickly.

Approach: Turba's approach uses available technologies that have already been demonstrated to be technically viable. This means they have eliminated much of the risk associated with new technology. However, they must establish internal expertise with Bluetooth technology, which does carry risk. The infrastructure must support the development of this expertise.

Infrastructure requirements: Turba will license cellular and Bluetooth technologies. They will hire hardware and software developers experienced in developing products using these technologies. Bluetooth engineers are highly sought after in the marketplace, so Turba needs to develop appropriate incentives to attract them.

Turba will also need to purchase the appropriate hardware, software, lab, machinery, and other equipment necessary for this new product category.

Turba's PERS Product Platform Strategy

Platform Strategy

The product platform will be established with the first PERS product and designed with future enhancements in mind (Figure 3.10). At a minimum, the platform will contain the form factor, battery size, and communication structure.

Turba's Multilevel Roadmap

Roadmaps

A multilevel roadmap shows the relationships between elements of the Product Tier. Turba's roadmaps are explained and illustrated in the next chapter (See Sidebar 4.1).

Figure 3.10: Turba's PERS Platform

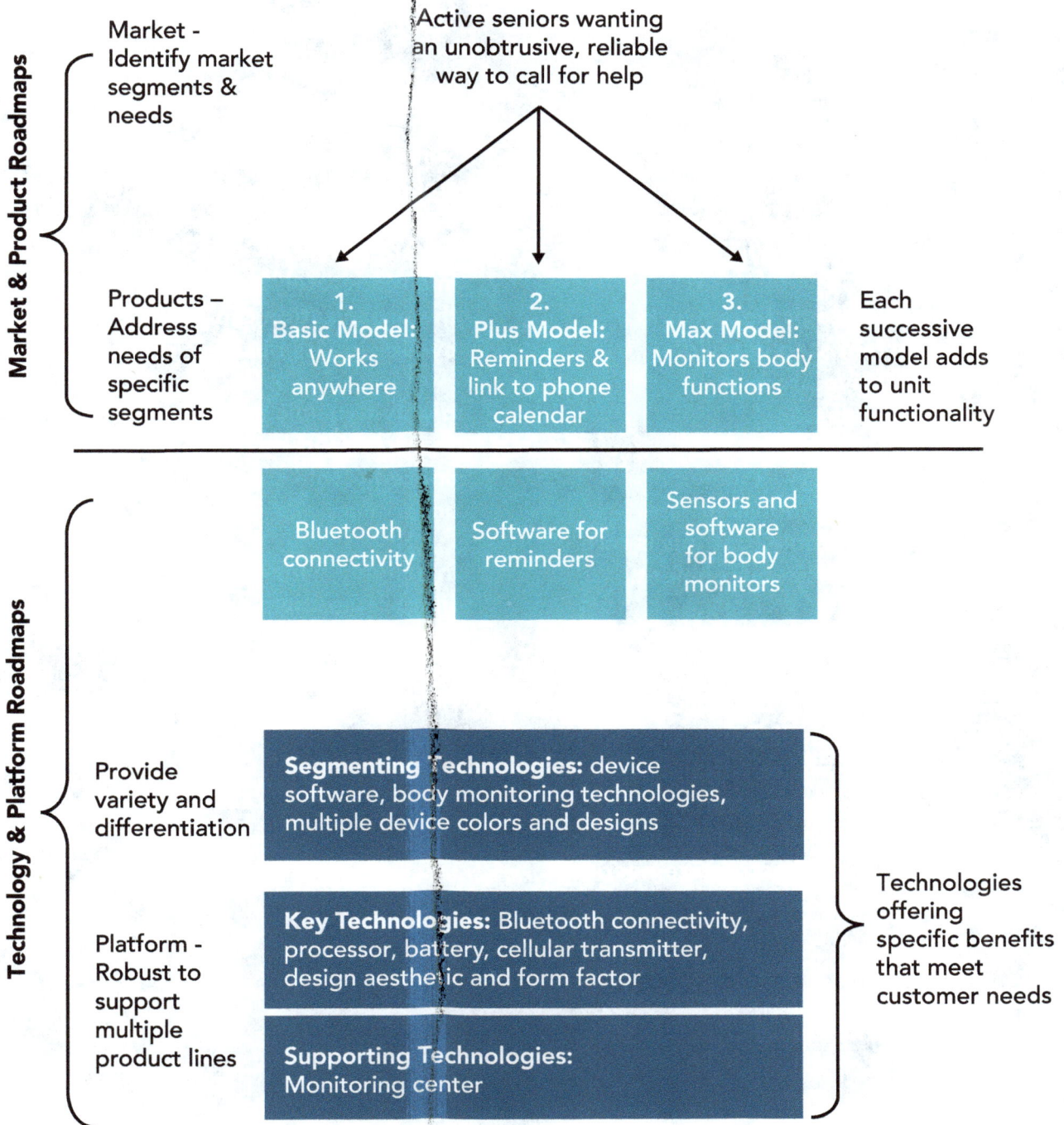

Market & Product Roadmaps

Market - Identify market segments & needs

Active seniors wanting an unobtrusive, reliable way to call for help

Products – Address needs of specific segments

1. Basic Model: Works anywhere
2. Plus Model: Reminders & link to phone calendar
3. Max Model: Monitors body functions

Each successive model adds to unit functionality

Bluetooth connectivity

Software for reminders

Sensors and software for body monitors

Technology & Platform Roadmaps

Provide variety and differentiation

Segmenting Technologies: device software, body monitoring technologies, multiple device colors and designs

Platform - Robust to support multiple product lines

Key Technologies: Bluetooth connectivity, processor, battery, cellular transmitter, design aesthetic and form factor

Supporting Technologies: Monitoring center

Technologies offering specific benefits that meet customer needs

Key Chapter Points

1. **Strategy shapes an innovation program by defining how product categories, technological capabilities, and other assets evolve.**

2. **Creating a strategic framework is an important step in building a more adaptive product development system.**

3. **Traditional strategy assumes that markets and industries are fundamentally stable, but this is rare in today's business environment.**

4. **The s2m Strategic Framework provides three different frameworks (Stable, Moderately Dynamic, and Highly Dynamic environments), covering the characteristics of each environment, a description of each strategic framework, and the internal capabilities and infrastructure needed for each model.**

5. **The strategic framework in a stable environment is relatively immobile, with little or no change.**

6. **Some enterprises act like they're in a stable environment when they are not because they do not recognize change.**

7. **In a moderately dynamic environment, the organization can adapt to change by altering its business model.**

8. **Highly dynamic environments are characterized by rapid, discontinuous, unpredictable changes, resulting in uncertainty and instability.**

9. **In a highly dynamic environment, simple rules help an organization capture unexpected opportunities amid market confusion.**

Notes

1. Michael Song and Yan Chen, "Organizational Attributes, Market Growth, and Product Innovation," *Journal of Product Innovation Management* 31(6) (2014): 1312–29.

2. Donald Sull, Rebecca Homkes, and Charles Sull, "Why Strategy Execution Unravels—and What to Do About It," *Harvard Business Review* (March 2015).

3. A classic approach to understanding an industry and how change can affect strategy is Michael Porter's book *Competitive Strategy: Techniques for Analyzing Industries and Competitors* (New York: Free Press, 1998). Chapter 1 describes five-forces analysis, and Chapter 8 describes industry evolution.

4. Christopher B. Bingham and Kathleen M. Eisenhardt, "Position, Leverage and Opportunity: A Typology of Strategic Logics Linking Resources with Competitive Advantage," *Managerial and Decision Economics* 29(2–3) (2008): 241–56, https://doi.org/10.1002/mde.1386.

5. Kathleen Eisenhardt and Donald Sull, "Strategy as Simple Rules," *Harvard Business Review* (January 2001).

6. This discussion is grounded in several articles, including Michael E. Porter, "What Is Strategy?," *Harvard Business Review*, November–December 1996; Kathleen M. Eisenhardt and Shona L. Brown, "Patching: Restitching Business Portfolios in Dynamic Markets," *Harvard Business Review* (May 1999), 72–82; Kathleen M. Eisenhardt and Jeffrey A. Martin, "Dynamic Capabilities: What Are They?," *Strategic Management Journal* 21(10–11) (October 2000): 1105–21, https://doi.org/10.1002/1097-0266(200010/11)21:10/113.0.co;2-e; Christopher B. Bingham and Kathleen M. Eisenhardt, "Position, Leverage and Opportunity: A Typology of Strategic Logics Linking Resources with Competitive Advantage," *Managerial and Decision Economics* 29(2–3) (2008): 241–56, https://doi.org/10.1002/mde.1386; and Martin Reeves, Ming Zeng, and Amin Venjara, "The Self-Tuning Enterprise," *Harvard Business Review* (June 2015), 76–83.

7. Steven Wheelwright and Kim Clark, *Revolutionizing Product Development*, (New York: The Free Press, 1992).

8. A practical explanation of business models and tools can be found in Alexander Osterwalder, Yves Pigneur, Tim Clark, and Alan Smith, *Business Model Generation: A Handbook for Visionaries, Game Changers, and Challengers* (Hoboken, NJ: Wiley, 2010).

9. Eisenhardt and Martin, "Dynamic Capabilities."

10. Zelong Wei, Yaqun Yi, and Hai Guo, "Organizational Learning Ambidexterity, Strategic Flexibility, and New Product Development," *Journal of Product Innovation Management* 31 (2014): 832–47, https://doi.org/10.1111/jpim.12126.

11. Ibid.

12. Ambidexterity is a very interesting topic. Although quite a bit of scholarly work is available, we recommend the following three sources: Julian Birkinshaw and Cristina Gibson, "Building Ambidexterity into an Organization," *MIT Sloan Management Review* (Summer 2004); Constantinos Markides and Wenyi Chu, "Innovation through Ambidexterity: How to Achieve the Ambidextrous Organization," London Business School, January 2009; Wei et al., "Organizational Learning Ambidexterity, Strategic Flexibility, and New Product Development."

13. D. Charles Galunic and Kathleen M. Eisenhardt, "Architectural Innovation and Modular Corporate Forms," *Academy of Management Journal* 44(6) (2001): 1229–49, https://doi.org/10.5465/3069398.

14. Alex Eule, "Are We There Yet? Self-Driving Has a Long Road Ahead," *Barron's*, May 3, 2021.

15. Eisenhardt and Sull, "Strategy as Simple Rules."

16. Christopher B. Bingham, and Kathleen M. Eisenhardt, "Rational Heuristics: the 'Simple Rules' That Strategists Learn from Process Experience," *Strategic Management Journal* 32(13) (2011): 1437–64, https://doi.org/10.1002/smj.965.

17. Kathleen Eisenhardt and D. Charles Galunic. "Coevolving: At Last, a Way to Make Synergies Work," *Harvard Business Review* (January 2000), 91–101.

18. Eisenhardt and Sull, "Strategy as Simple Rules."

19. Shona L. Brown and Kathleen M. Eisenhardt, "The Art of Continuous Change: Linking Complexity Theory and Time-Paced Evolution in Relentlessly Shifting Organizations," *Administrative Science Quarterly* 42(1) (March 1997): 1–34, https://doi.org/10.2307/2393807.

20. Eisenhardt and Sull, "Strategy as Simple Rules."

21. Kathleen M. Eisenhardt and Henning Piezunka. "Complexity and Corporate Strategy," in *The Sage Handbook of Complexity and Management*, edited by Peter Allen, Steve Maguire, and Bill McKelvey, 506–23 (Los Angeles: Sage, 2011), https://static1.squarespace.com/static/50d63bc4e4b0e383f5b2a05a/.

22. Brown and Eisenhardt. "The Art of Continuous Change."

Segment 2

Ideas and Selection

Chapter 4
Idea Management System

Chapter 4 Contents

What to Expect

The Ideas & Selection segment provides a structure for ensuring the organization maintains a steady supply of new product ideas. Ideas continuously and systematically generated from various sources provide fuel for organic growth. This chapter will describe how the idea management system is integral to ExPD.

We provide an overview of the ideation model that includes generating, capturing, preparing, and prioritizing (PV1) product ideas (Figure 4.1).

Figure 4.1: Ideas & Selection Segment

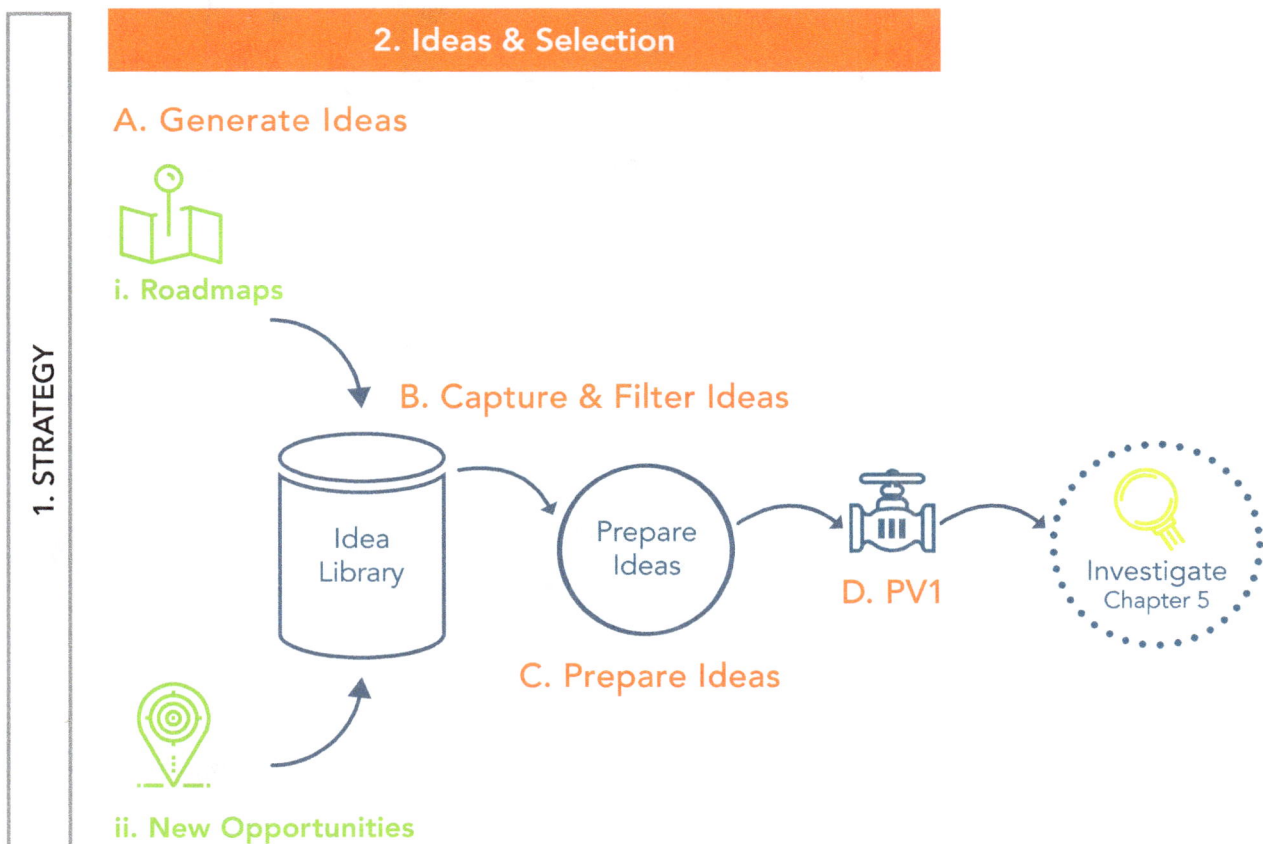

Most companies have some variation of the ideation process. ExPD has two entry points for generating ideas, including ideas on a roadmap or new opportunities from internal and external sources.

Product ideas are captured and filtered in the Idea Library. Many organizations and consultants talk about data "repositories." We use the term *library* instead because we are proponents of the team actively checking ideas in and out of the library. A repository is just for storage; it doesn't carry that connotation of actively using the ideas.

We describe a method to manage the Idea Library and to scale the quantity and flow of ideas. We then summarize the process of preparing the idea with the necessary information for the management team to decide on a product idea.

Next, we introduce the first Prioritization Valve (PV1) within the ExPD process. Before an idea enters PV1, the management committee assesses whether the product idea will proceed to the next step in the process. If the idea is chosen to proceed, it enters PV1, and the inner working of the valve begins, including scoring, selecting, and prioritizing incoming ideas. We use the image of a valve to illustrate controlling the rate of idea flow based on the available resources.

We conclude the chapter with a closer look at the management committee responsible for prioritizing ideas. We consider the composition of the committee, the advantages of early involvement in the process, and some practices to reduce bias in decision-making.

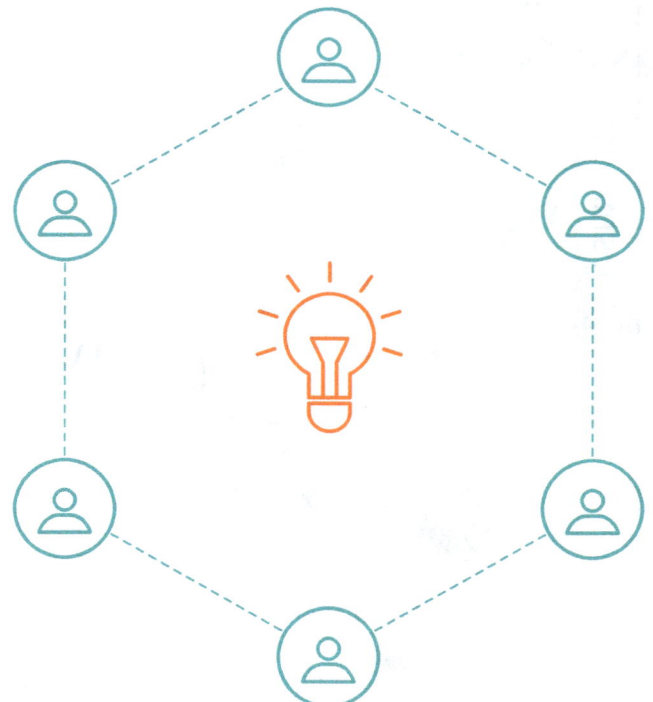

Overcoming the Hurdles of the Idea Management System

Most enterprises do a respectable job of handling new ideas, but we sometimes encounter an organization with no formal, structured idea management system. Often, ideas are not generated systematically, with well-designed frames, filters, and team participation, nor are they consistently scored, selected, prioritized, or collected in one central library.

The lack of a systematic approach can result in a variety of unproductive outcomes. Table 4.1 lists some of the common issues we have experienced with idea management systems. As we have encountered these, we have established ways to address them with ExPD, as summarized in the right column of Table 4.1. The following description of the Ideas & Selection segment shows how these methods can overcome the common hurdles in the idea management system.

Table 4.1. How ExPD Addresses Common Idea Management Issues

Idea Management Issues	ExPD Methods to Address the Issue
Gaps in the supply of ideas that fit strategic needs and opportunities	Product Roadmaps Ideation Engine
Revisiting the same ideas multiple times	Idea Library
Losing valuable ideas and preparatory work, with the potential loss of intellectual property	Idea Library
Placing too many ideas without enough resources into the product development process	Prioritization Valve (PV1)

Segment 2: Ideas & Selection

The Ideas & Selection segment within ExPD consists of four key areas: generate ideas, capture and filter ideas, prepare ideas, and Prioritization Valve 1 (Figure 4.2). Let's look at the ExPD process in more detail, starting with idea generation, when ideas enter the process.

Figure 4.2: Generate Ideas (i, Roadmaps, ii. New Opportunities)

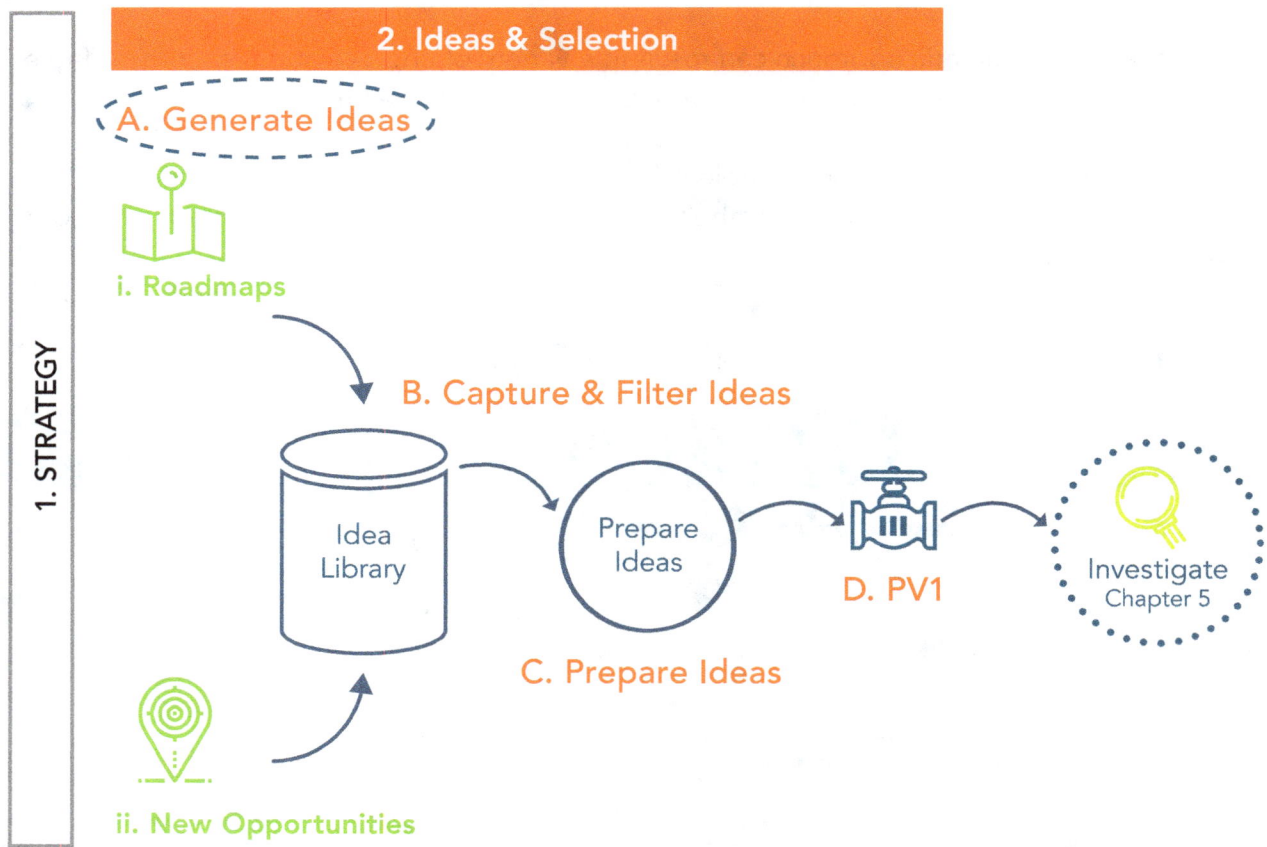

Generate Ideas

Ideas enter the Ideas & Selection segment through two entry points: ideas based on strategy and product roadmaps, and ideas based on new opportunities identified through other activities, including conversations with customers and partners, market scans, employee ideation contests, brainstorming sessions, open innovation methods, and employee submission forms.

Typically, at companies where people complain about not having enough good ideas, we find a lack of a systematic and balanced approach that leverages product roadmaps. We are proponents of shifting from solely relying on serendipitous ideas to focusing on a strategic approach that requires upfront planning.

Idea Generation via the Strategic Framework: Product Roadmaps

As noted in the introduction to this chapter, ExPD differs from most companies' ideation processes in having an integrated strategic framework via roadmaps. Product roadmaps show prioritized initiatives resulting from strategic planning typically executed by product management and the project team. A roadmap can focus on a single market, product, technology, or platform strategy, but it can be illuminating to present related strategies on a multilevel roadmap[1] (Figure 4.3).

Figure 4.3: Multilevel Roadmap

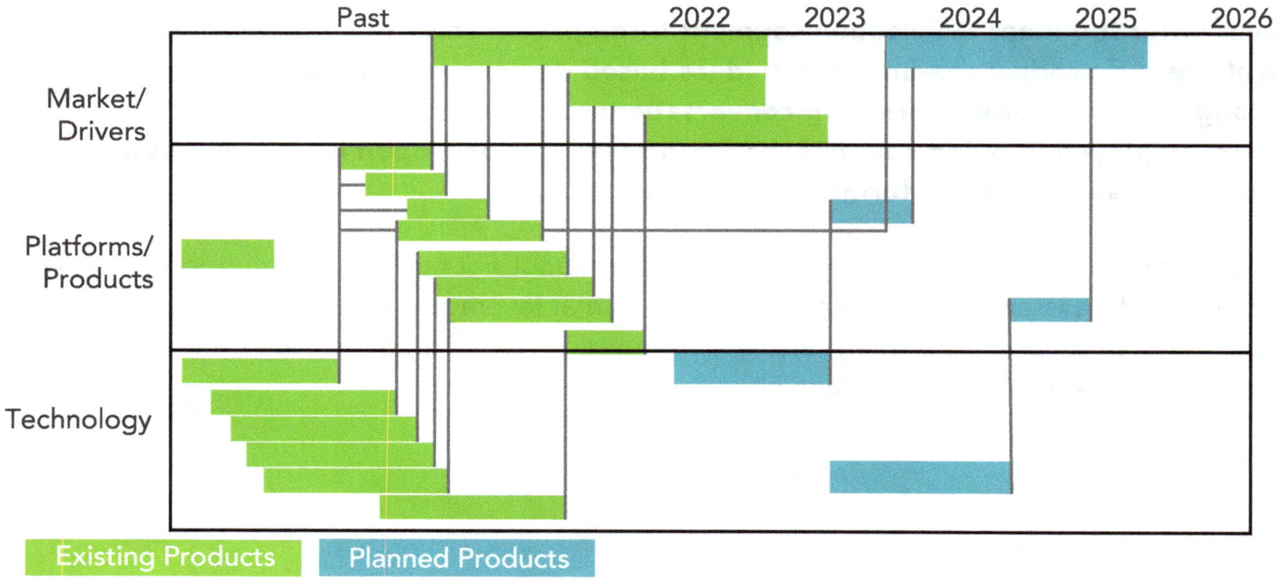

Roadmaps are a graphical representation of the sequence and timing of initiatives or projects, showing time along a horizontal axis. Considerable research and thought go into developing the roadmaps, so these ideas should proceed through the idea management process with few obstacles.

We have heard confusing information from different practitioners and consultants that product roadmaps are a strategy. We want to clarify that roadmaps are not a strategy but rather a graphical representation of product strategies. Unfortunately, some practitioners race to build roadmaps without the necessary product strategies to back them up.

Unfortunately, some practitioners race to build roadmaps without the necessary product strategies to back them up.

When done correctly, roadmaps are particularly useful in the widespread communication of organizational priorities. They also support the forecasting of skills and technologies to be acquired to meet strategic goals. Turba worked as a cross-functional team in developing its multilevel roadmap (see Sidebar 4.1).

Sidebar 4.1. Turba's Multilevel Roadmap

TURBA CORPORATION

A Turba team that included product management, design, engineering, and operations met to construct a multilevel roadmap. The team was partial to the multilevel format since it showed the relationship between elements of the product strategies (Figure 4.4).

Figure 4.4: Turba's Multi-level Roadmap

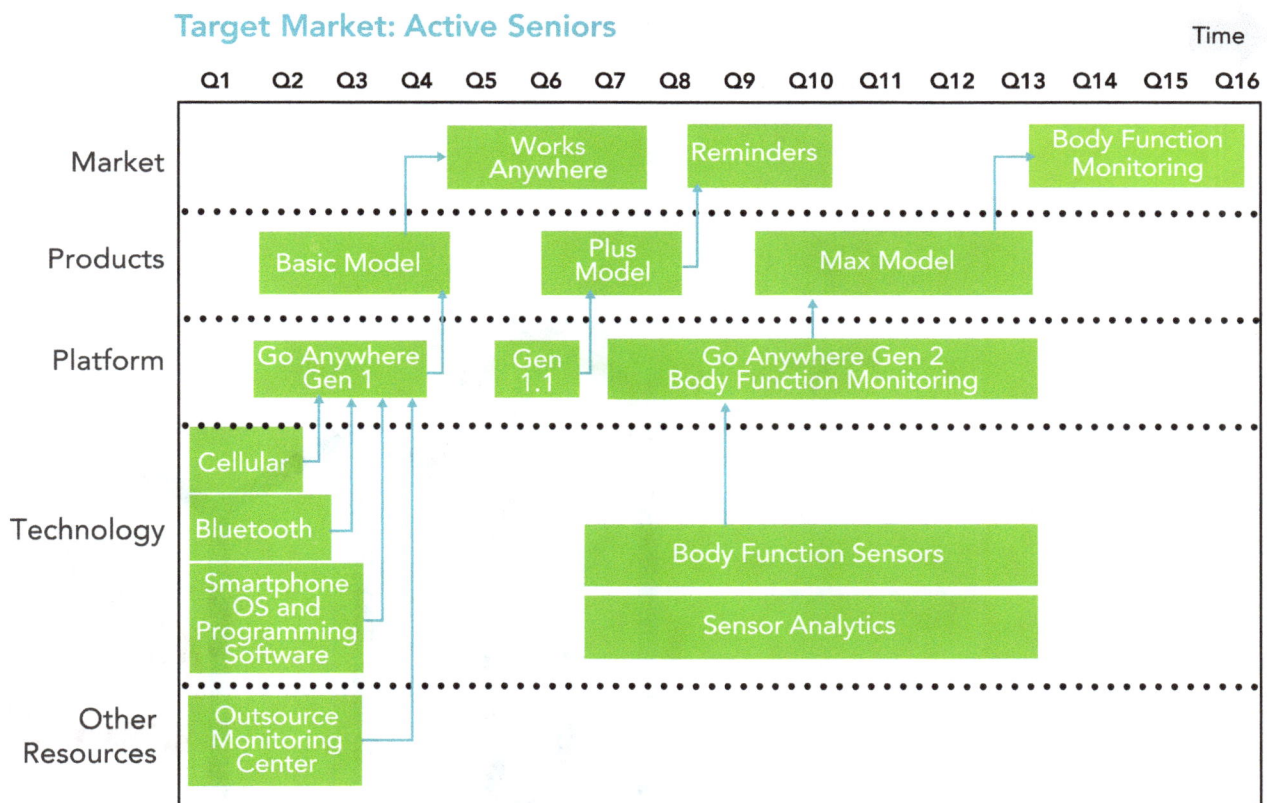

The roadmap represented five different levels (Market, Products, Platform, Technology, and Other Resources). Each level required an estimation of the time needed to execute the strategy, as shown across the top of the roadmap, and identification of dependencies between the levels, shown by arrows connecting the boxes.

Market reflects the market strategy and opportunities to satisfy customer needs. These represent the "whys" of the roadmap. **Products** reflect product strategy, representing the "whats" of the roadmap—what will be created to meet each opportunity in the market roadmap. To deliver on the "whats" requires the **platform** and **technology** strategies, plus **other resources**, such as new distribution channels, suppliers, and manufacturing processes. These represent the "hows" of the roadmap and can be displayed in multiple layers below the product level.

Building dependencies and timing into the roadmap allowed the team to see obstacles to meeting the target dates. Based on these insights, the Turba team acknowledged that it might be necessary to rework some of the tactics underlying the strategies.

Overall, the management team was pleased with the coordinated outcome of the multilevel roadmap since it reflected the product strategies and it enabled buy-in across the disciplines.

The term *roadmap* sometimes encourages enterprises to treat these documents as fixed and difficult to change. It is essential, however, to balance being rudderless and constantly changing with being structured and rigid. Suppose your enterprise is in a highly dynamic environment. In that case, we recommend that roadmaps be broadly directional with shorter time horizons (three to six months). In contrast, traditional roadmaps are narrow and highly structured, with longer time horizons of two to three years.

In a highly dynamic environment, roadmaps that are broadly directional with shorter time horizons are appropriate.

Unfortunately, some companies are unwilling to invest resources in the necessary research and planning that roadmaps need. In some cases, we find a lack of cross-functional integration between engineering and marketing in roadmap development. It is not unusual to see engineering and marketing work on their separate roadmaps. To drive the business in a unified direction, they must collaborate on, among other things, monitoring new technology developments and external market forces. Their roles as partners are significant for keeping roadmaps robust, up-to-date, and adaptable.

Idea Generation via New Opportunities

Along with roadmaps, organizations have alternative ways for ideas to enter the process. Ideation is the critical bridge between strategy and development. Ensuring enough of the right ideas requires a broad-based, diverse, robust ideation process, driven by strategy and generating outputs appropriate to your development process, something we call an "Idea Engine."

Ideation is the critical bridge between strategy and development.

Ideally, your Idea Engine should have the following six characteristics:

1. It should be highly dynamic and proactive (continuously generating new possibilities for your development teams), as well as adaptive (quickly shifting focus and priorities to adjust to tactical and strategic imperatives).

2. It should fully leverage internal communities, including employees in relevant roles and divisions across the enterprise, and external communities and resources, such as customers, partners, and industry experts.

3. It should work toward a balanced portfolio by addressing the need for breakthrough or new-to-the-company product ideas, as well as more frequent and focused ideas to help improve a feature or function of an existing product.

4. It should include various techniques, sources, and participants, to ensure you have truly comprehensive coverage of the idea space relevant to your company.

5. It should include the systematic scanning of your ecosystem, market, and related industries and fields to identify potential ideas.

6. It should be designed specifically for your enterprise, culture, ecosystem, and strategy.

The Ideation Model

At its most basic, ideation is a two-step process of divergence and convergence (Figure 4.5). The team starts with a challenge, problem, or opportunity. The goal is to finish with one or a few potential solutions or concepts. In between, you want a large pool of ideas representing a significant cross-section or sampling of the options (product or service concepts, solutions, opportunities, etc.) in a particular idea space (the field of all possible ideas related to the challenge, problem, or opportunity).

Figure 4.5: Divergence & Convergence

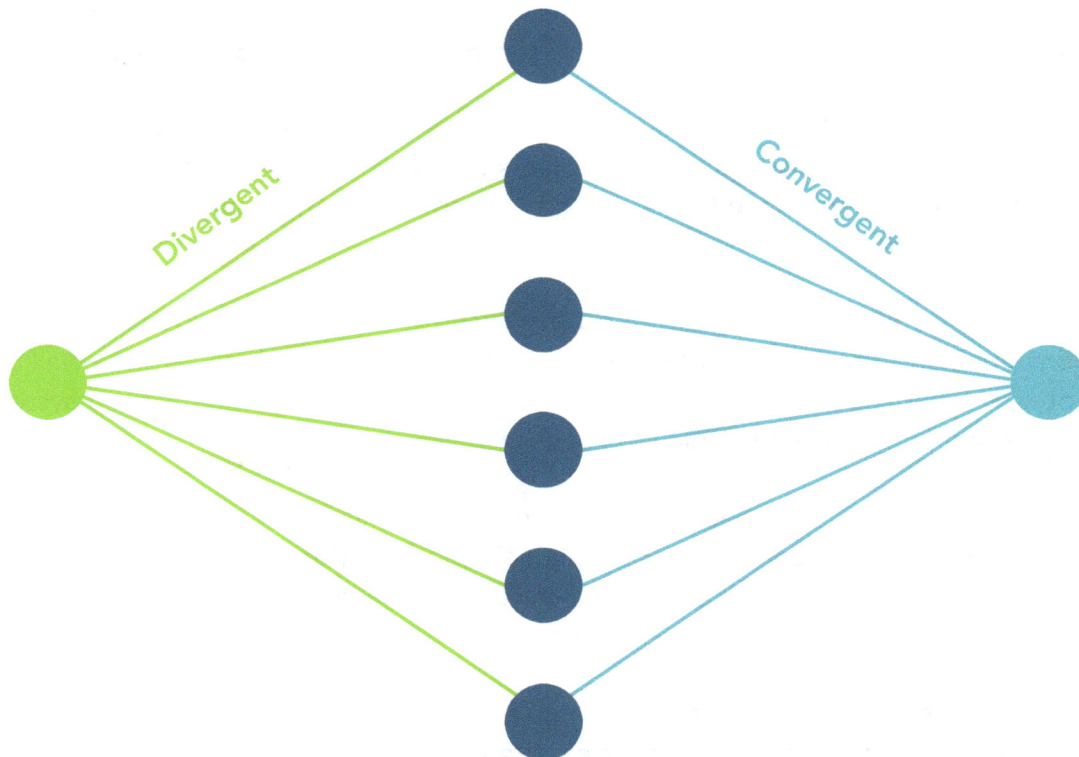

In general, the team is looking for more/different ideas. A greater quantity of ideas increases the probability of discovering the best possible idea. Convergence involves taking the ideas collected during your divergence session and narrowing them down to a select few solutions.

Designing an Idea Engine

To operationalize the model, we define some functional elements. An ideation session is a single iteration of the process—for example, an exercise, workshop, or time-limited initiative to develop a defined set of ideas. Each session is characterized by the following five activities:

1. The **input method** is the mechanism for generating ideas. Typical methods include brainstorming, innovation challenges, suggestion boxes, customer advisory panels, market scans, and ethnography studies.

2. The **frame** is a statement describing the problem, opportunity, or challenge; it delineates the boundaries of an appropriate idea. In this way, the frame defines the boundaries of the idea space. For example, a team might be charged with identifying ideas that serve a particular customer group or apply a specific promising technology.

3. The **filter** defines the relative value of the idea—for example, criteria, metrics, or priorities. The team might filter out projects that are not forecast to deliver a set return on investment, have too large of a carbon footprint, or can't be developed within two years.

4. The **collection** function is the method by which the team gathers ideas. It can range from small and low-tech to cutting-edge technology applications—from a wall of sticky notes in a single session to a sophisticated database or idea management system that aggregates ideas from multiple sessions.

5. The **selection** function is the process of choosing ideas to prepare for consideration. It may be as straightforward as applying a simple filter (for instance, a voting exercise) in a single session, or it can leverage a series of sophisticated filters across multiple sessions, using advanced scoring methods.

The Idea Engine model operates as shown in Figure 4.6. Using the frame that identifies a problem, opportunity, challenge, or seed idea, the team collects new ideas, filters them, and selects a few for consideration.

Figure 4.6: Idea Engine

Generate:
Brainstorm with internal teams

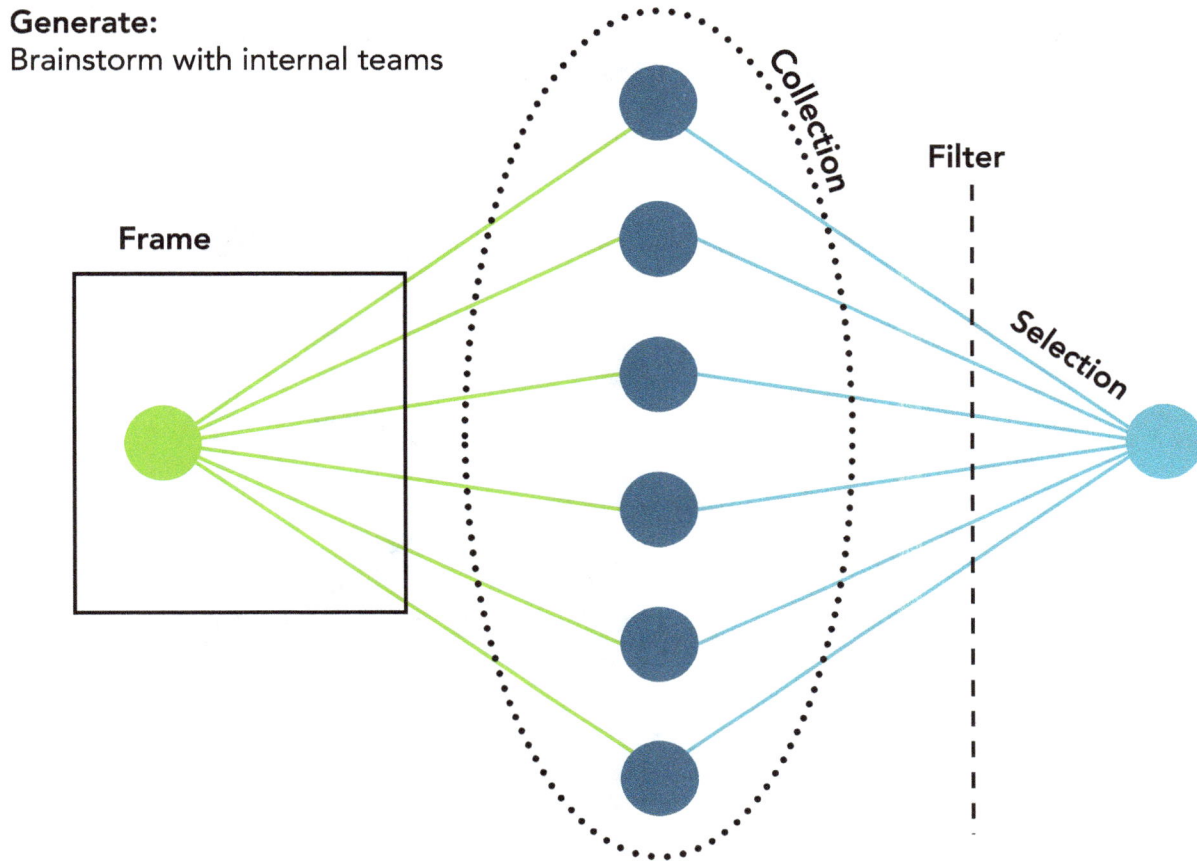

In Figure 4.6, there is one source of ideas, brainstorming with internal teams; however, this is only one of the possible input methods.

Another is to invite a relevant community—employees, customers, or partners—to submit ideas, either formally or informally (Figure 4.7). Here the frame is even more critical, and you need to make it very clear what kinds of ideas you are asking for. The collection and selection phases also may need to be more explicit and transparent.

Figure 4.7: Critical Frame Requirements - Community

Solicit:
Ask a relevant community

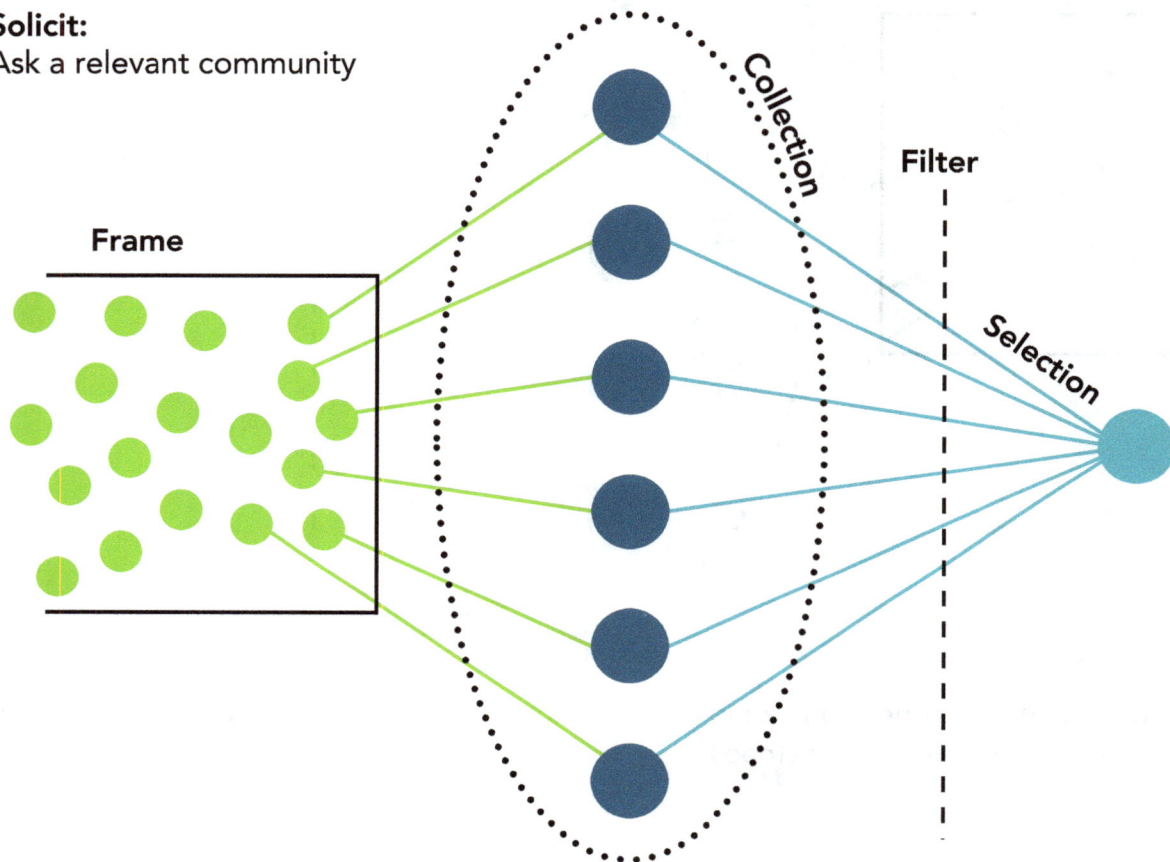

At the broadest level, some companies solicit thoughts from thousands of employees and customers, harvesting thousands of ideas as input and using a range of online idea management tools to enable the collection and selection process. You may want to cultivate an idea-rich relationship with a specific community beyond a one-off session of this type, encouraging employees, partners, and customers to submit ideas regularly and even participate in brainstorming sessions.

Also, you can proactively look outside your enterprise and immediate communities through a scan of relevant markets and industries (Figure 4.8). Here, your team members are looking for existing ideas that can be adapted.

Figure 4.8: Critical Frame Requirements – Relevant Markets and Industries

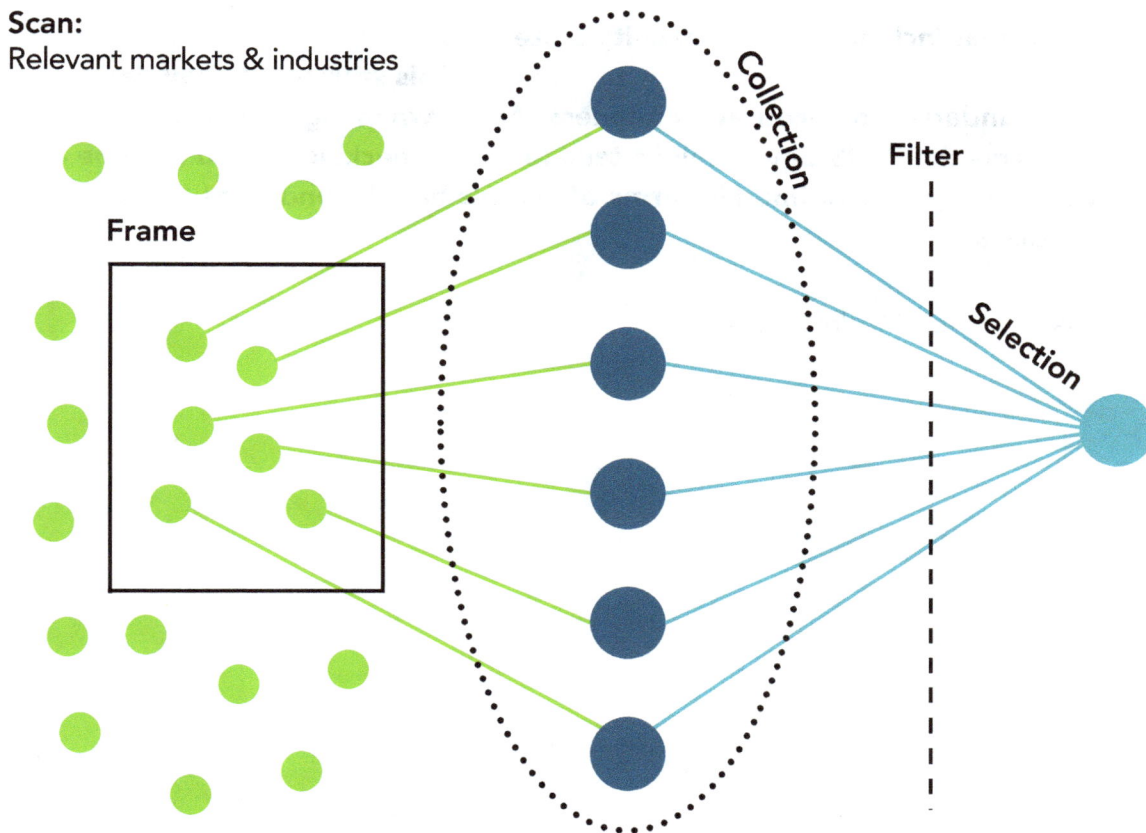

Scan:
Relevant markets & industries

Frame

Collection

Filter

Selection

The challenge with this approach is finding the right area to scan. Looking too close to your existing markets may just turn up those "usual suspect" ideas that your competitors are already exploiting or have already eliminated for good reasons. Looking too far afield risks being irrelevant and wasting time. But great ideas are often inspired by concepts from very different fields, and a seemingly impossible concept is often a good seed idea to encourage a brainstorming session. Sidebar 4.2. provides a Turba example outlining the process of using a frame during a brainstorming session to provide focus.

Looking too close to your existing markets may just turn up those "usual suspect" ideas that your competitors are already exploiting or have already eliminated for good reasons.

Sidebar 4.2. Turba's Idea Engine

TURBA CORPORATION

Turba used a frame that included an opportunity statement: There is a need for an emergency alert system outside the home for the elderly. **This statement provided focus and dimensional boundaries to encourage openness. As shown in Figure 4.9, this frame included several forms the PERS device might take (a watch, necklace, or some type of clothing) and customer groups defined in terms of chronic health conditions (dementia, diabetes, and cardiac).**

Figure 4.9: Turba's Ideation Model Frame

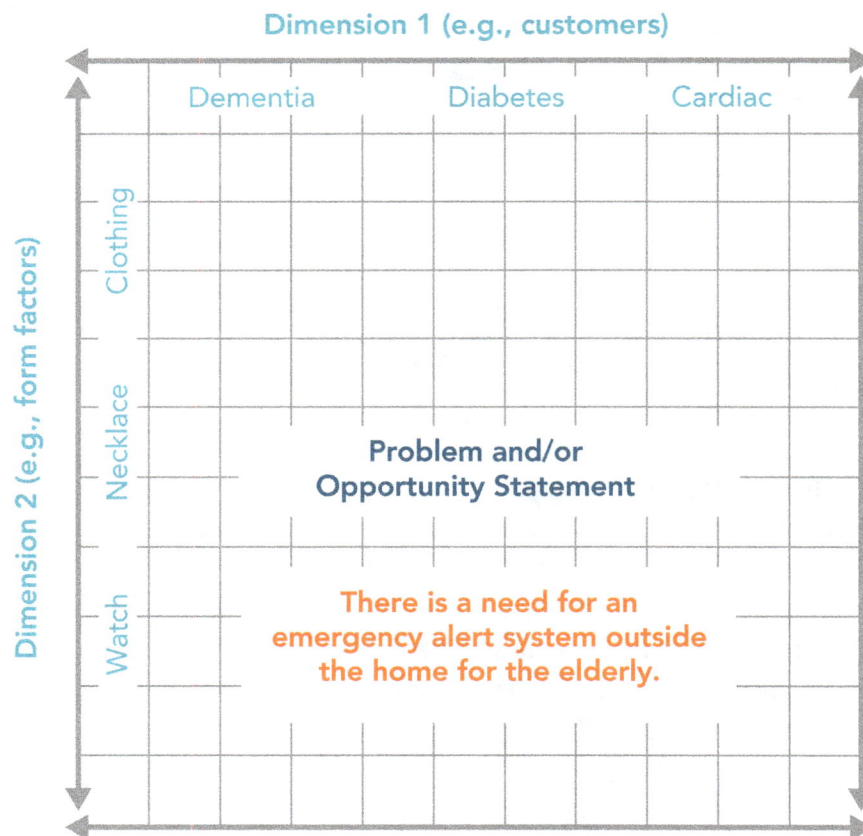

Dimension 1 (e.g., customers)

Dementia   Diabetes   Cardiac

Dimension 2 (e.g., form factors)

Clothing   Necklace   Watch

Problem and/or Opportunity Statement

There is a need for an emergency alert system outside the home for the elderly.

Within this frame, Turba could either go broad and look at all the form factors and customer types or zero in on a particular area of the frame and design a new frame for brainstorming more specific ideas. Using a frame is like taking pictures with a zoom lens; it allows the team to define the ideation effort very broadly or very specifically.

After defining the frame for this round of idea generation, the team defined filters (criteria, metrics, and priorities). This effort enabled the team to select the best ideas generated to feed into the pipeline.

A significant advantage of emphasizing a proactive and standardized Idea Engine based on framing opportunities was that Turba obtained a pool of high-quality ideas that can move quickly through the ExPD process.

Note that Turba didn't build their Idea Engine all at once. They developed it through an iterative process, testing different activities and methods. They developed a robust idea-generating process and culture over a series of projects.

For the PERS project, the team began with a brainstorming session focused on elderly persons falling due to dementia, diabetes, and cardiac issues. The form factor dimension focused on the watch.

Then they conducted a market scan of related products and services and an idea contest with employees and partners. The resulting ideas and materials were cataloged in a web tool, and the collection and selection processes were documented and turned into a workbook for innovation teams. Further initiatives and sessions expanded the focus of ideation to additional conditions, users, and scenarios. The company expanded the Idea Library, which became an online resource available across the business.

There are many unique styles and flavors of ideation. Still, this basic descriptive model applies whether ideation involves a few people brainstorming in a room or a company-wide challenge.

Capture and Filter Ideas

Do you hear the same ideas repeatedly? This may be a symptom of not having an Idea Library to capture and store ideas that enter the process (Figure 4.10). Often, during ideation meetings, participants say, "Didn't we propose something like this a couple of years ago? Whatever happened to that idea?"

Figure 4.10: Capture & Filter Ideas

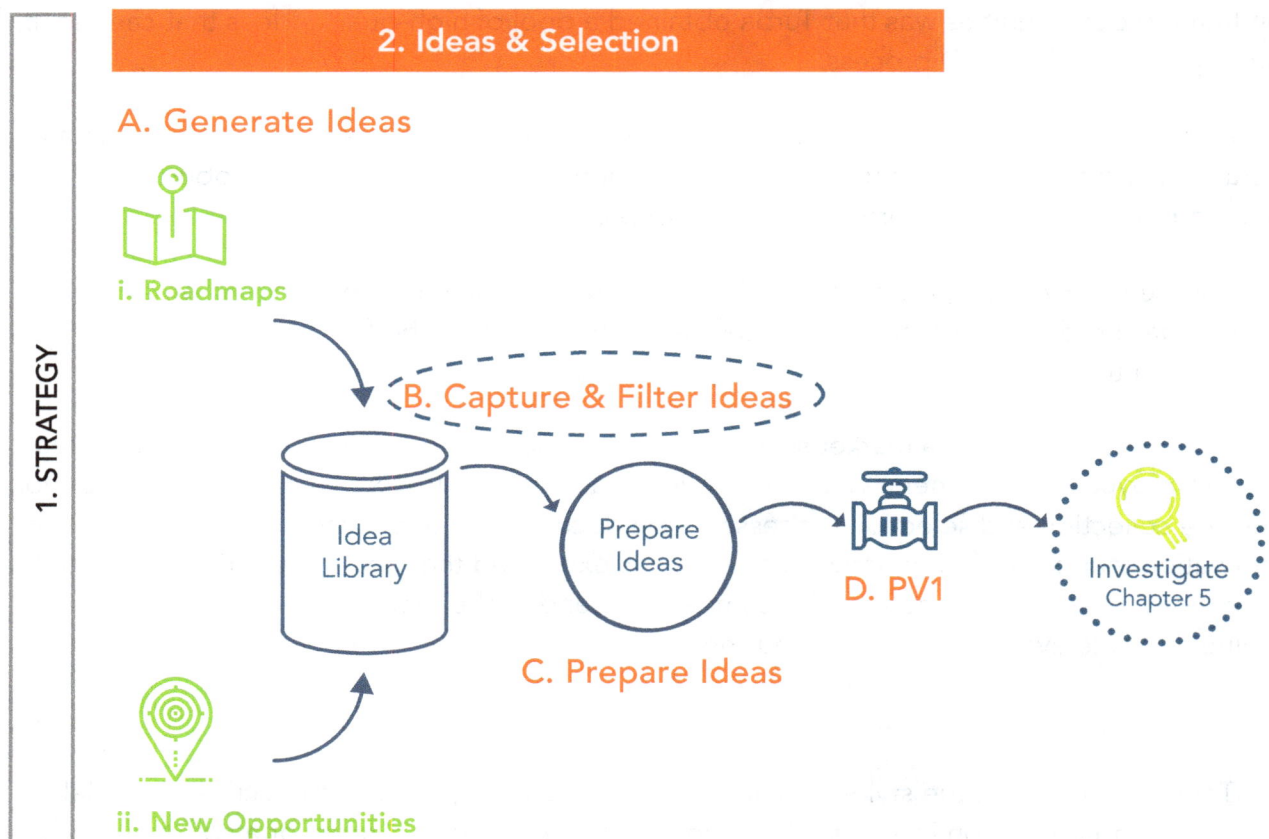

Do you hear the same ideas repeatedly? This may be a symptom of not having an Idea Library to capture and store ideas that enter the process.

Capturing Ideas in the Idea Library

The Idea Library is a critical component of your Idea Engine. It includes whatever descriptive material was associated with each product idea and the applied selection criteria. Even if the idea was rejected, this data is still useful; it lets you determine why the idea was put on hold or canceled and perhaps revive the idea if appropriate.

Usually, a substantial amount of work goes into determining the disposition of the idea; you don't want to lose or replicate all that work. An Idea Library can also help if you're changing strategy because of a change in the marketplace or other factors. One of the first things you might do is review rejected or inactive ideas to see if they are now relevant from a new perspective.

Ideally, all the valuable ideas that enter an enterprise should be collected and organized in this way. All too often, companies lose ideas because the material is squirreled away on someone's hard drive or the person with the idea has left the company. This can also result in losses of important intellectual property.

All too often, companies lose ideas because the material is squirreled away on someone's hard drive or the person with the idea has left the company.

An Idea Library should be easy to use and accessible so that employees get in the habit of searching for and contributing ideas as part of their innovation, risk resolution, and problem-solving initiatives.

A library for capturing ideas is one of the easiest tools that an enterprise can implement. Initially, it can be as simple as capturing ideas in a spreadsheet or some collaborative workspace. Once you figure out the best process for capturing ideas, you can always progress to a more robust technology solution.

A library for capturing ideas is one of the easiest tools that an enterprise can implement.

Ideally, an idea is stored along with whatever work has already been done to qualify or develop the idea. For example, the frame statements and filter criteria are useful metadata to help you search for relevant ideas in the library or to establish rules for revisiting ideas under future scenarios. Without the metadata, the idea is less easily retrieved or re-analyzed. For an idea that has gone through a selection phase, a short list of associated data might include the following items:

1. Description

2. Frame (e.g., framing question, problem, or opportunity statement)

3. Filter (e.g., metrics, criteria, prioritization, or voting method)

4. Final rating based on filter

If a set of ideas resulted from an ideation session, you also want to document the session's overall goals and any relevant research or observations. When qualified ideas or sets of ideas are entered into the library, the process owner's responsibility is to track them and take the necessary next steps.

Filtering Ideas

Enterprises often employ different frames and filters to screen various ideas for different goals. It's possible to postpone selection and dump vast numbers of "raw" ideas into the pipeline, but we don't recommend that. Although each session should include a step where all ideas are welcome and recorded, it's better to do some qualifying and prioritization while the ideas are still fresh in employee's minds.

Note that generating more ideas rather than fewer has clear benefits; it opens up more possibilities, encourages a broader vision, and increases the likelihood that your ideas will include the best possible options to match your strategies. At the same time, there is an opportunity cost of disqualifying an idea too early.

As with any process, it's helpful to scale the quantity and flow of ideas generated to match your needs without overwhelming the resources of your innovation teams or compromising the integrity of decision-making processes. For example, Prioritization Valve 1 (PV1), described later in this chapter, is a rigorous selection session with an experienced cross-functional management committee that carefully looks at each idea submitted.

A level of selection needs to be completed before PV1. Ideally, there is a cadence of generation and selection sessions with target quantities at each step along the way. We call this "idea flow;" the optimal flow will, of course, differ depending on your resources, number of teams, areas of opportunity, and the goals of your innovation program.

Ideally, there is a cadence of generation and selection sessions with target quantities at each step along the way.

For example, suppose we started an ideation session with one idea, generated 20 ideas, and then selected the top 3. The flow would look something like this:

1 < 20 > 3

A series of five such sessions can be planned, each focusing on a different strategic opportunity area. A final selection session is held to pick one top idea in each area. The aggregated flow is represented as the following notation:

5 < 100 > 5

You can use this notation to discuss goals for your idea management process. How many ideas does the enterprise need to generate or collect, in which areas? How many ideas need to be selected at each step? What is the optimum number of ideas to feed into the process?

Idea flow is an important concept to guide the team in creating the optimal number of relevant product ideas, so the project team is positioned to work on the best ideas in the queue. The idea flow can be achieved by having an ongoing quarterly or semiannual brainstorming event.

Prepare Ideas

The third key area of the Ideas & Selection segment is Prepare Ideas. If an idea is on the roadmap, it has previously been vetted, so we know the idea is aligned with the strategy, and management has prioritized it in the portfolio. No further work is needed before moving to PV1, where a product idea transitions into a "project."

PV1 is where a product idea transitions into a "project" since resources are assigned to fleshing out the product idea.

Ideas not on the roadmap require more vetting for strategic fit. In some cases, the management committee will not have enough information to do a thorough assessment, and the project team needs to prepare a description of the idea (Figure 4.11).

Figure 4.11: Prepare Ideas

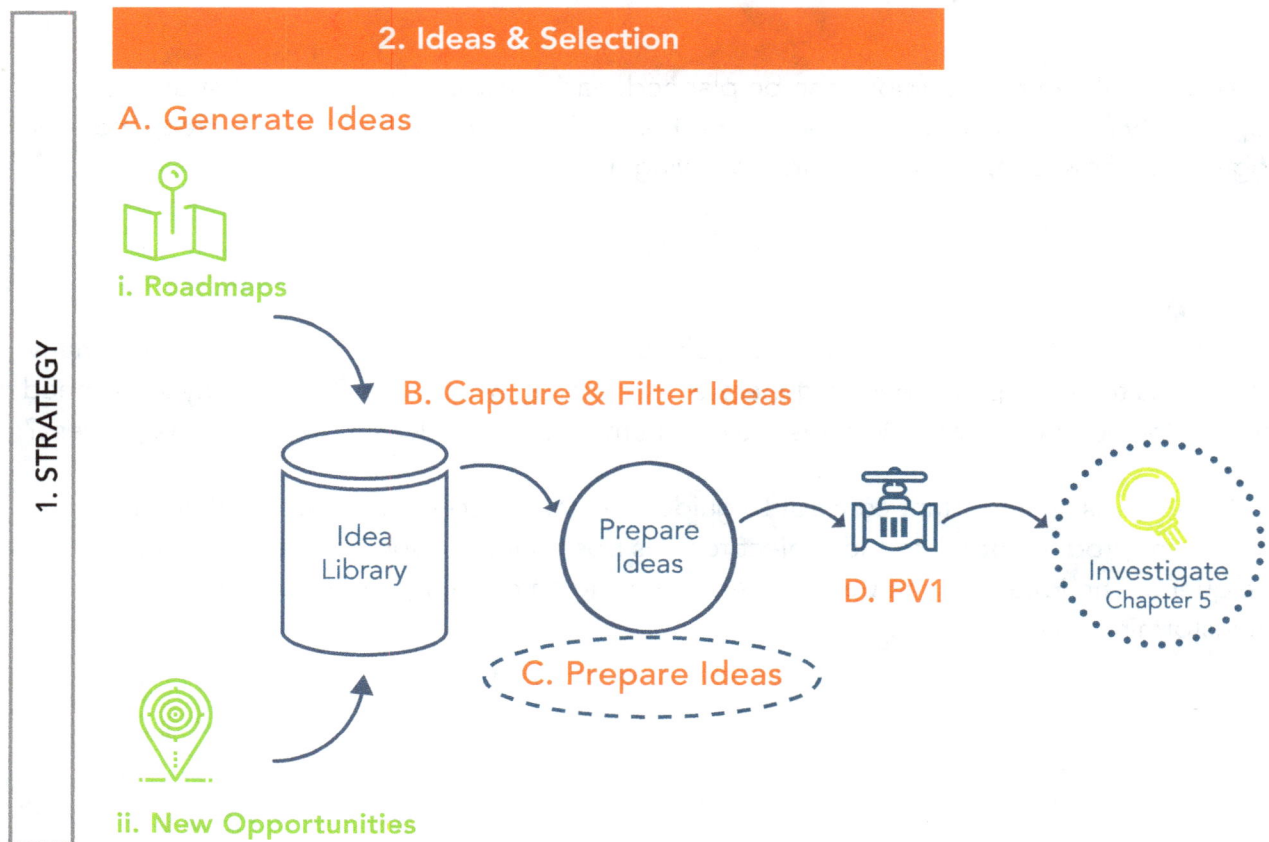

Typically, this involves answering the following six high-level questions for the management committee:

1. Is there a compelling reason to do this project?

2. Is the idea aligned with the business and its strategy?

3. Is there a competing product?

4. Is the competitive landscape favorable?

5. Does the idea meet an important customer need?

6. What are the critical next steps?

Usually, the project team can obtain this information using secondary sources, such as published materials and market studies, and primary sources such as observational techniques, and very early-stage prototyping.

Some management committee members want to have all the answers upfront when it is just an initial idea. For example: What is the ROI? How profitable will the product be? How much will it cost? The management committee must recognize that it is too early in the process to determine this type of detailed financial information, since the product idea is not properly fleshed out.

The management committee must be patient and have a tolerance for uncertainty.

Prioritization Valve 1 (PV1)

After the project team prepares a description of the idea, the management committee decides at PV1 whether the product idea should proceed to the next step in the process, Investigate (Figure 4.12). This assessment includes all incoming ideas: roadmap and new opportunities. The output is a decision made on each product idea.

Figure 4.12: Prioritization Valve 1 (PV1)

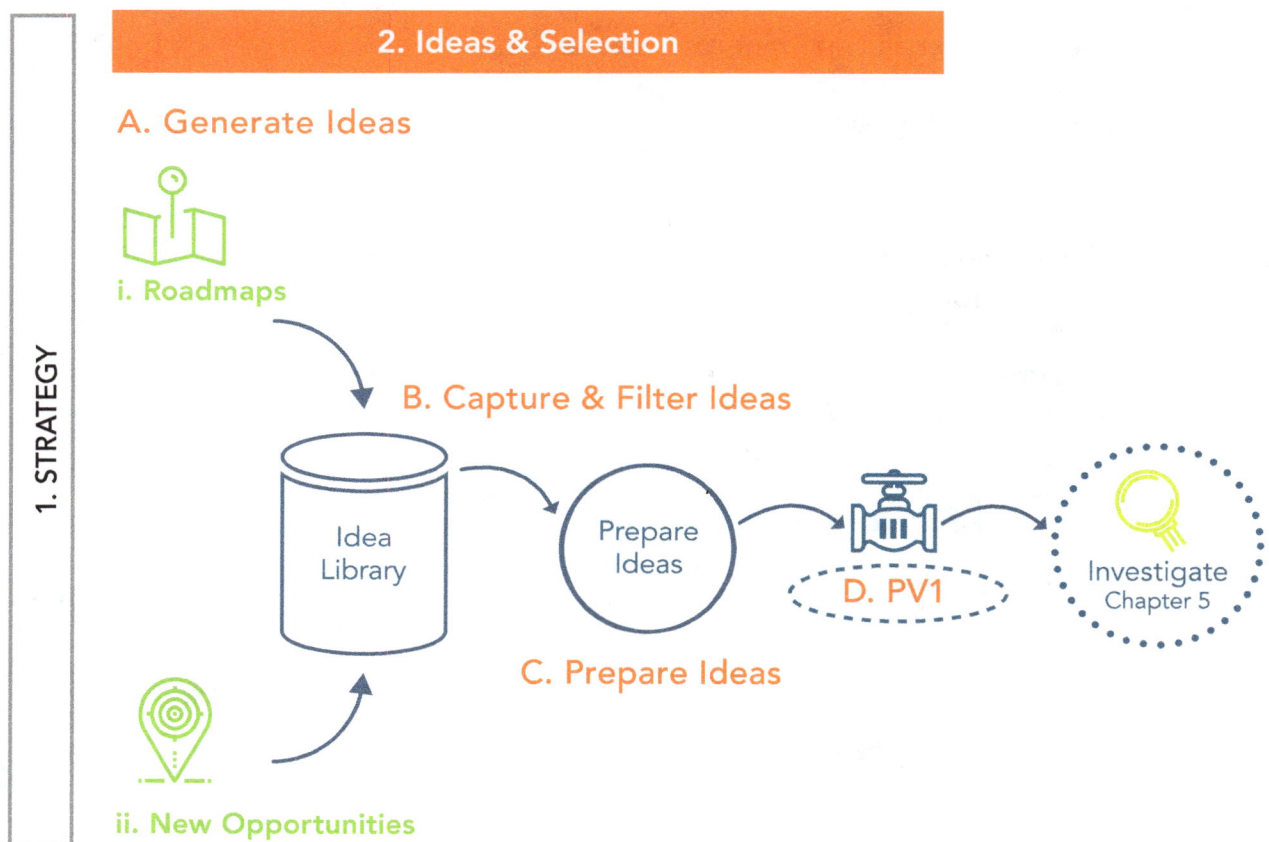

If the idea proceeds to PV1, the inner workings of the valve begins, including assessing, scoring, selecting, and prioritizing incoming ideas (Figure 4.13). If the committee believes an idea has promise and is willing to allocate time and resources, they can prioritize it within PV1. A valve is used to illustrate adjusting the rate of idea flow based on available resources. The valve can also have a trickle of ideas, smaller projects when the pipeline is busy, up to full bore when the pipeline has excess capacity.

Figure 4.13: Inner Workings of the Prioritization Valve

Individual ideas/concepts

A valve is used to illustrate adjusting the rate of idea flow based on available resources.

The prioritized ideas, represented by the individual cells inside the valve (Figure 4.14), are waiting for their turn to be assigned and released into development. Without this valve, the development pipeline is easily overloaded, and innovation can slow to a crawl. If the idea proceeds in the process, it enters the next step of the ExPD process, Investigate, discussed in the next chapter.

Figure 4.14: Projects in Queue

Selected and prioritized projects in the queue

Without this valve, the development pipeline is easily overloaded, and innovation can slow to a crawl.

We have witnessed many scenarios where management keeps squeezing in one more favorite project, regardless of whether resources are available or other projects will be adversely affected. ExPD addresses this issue by inserting prioritization valves throughout the process. If the management committee decides to continue pursuing an idea, they determine how it ranks against other ideas in the valve. If available resources are inadequate, the committee should place the idea in a queue until resources are available.

We have seen prioritization work well for our clients, following our top nine guidelines for managing resources:

1. Score and prioritize your most valuable ideas.

2. Staff the appropriate projects based on available resources; do not exceed available capacity.

3. New projects do not preempt projects in progress unless a significant business issue, such as a regulatory issue, is driving a priority.

4. New projects do not start until resources are available.

5. New projects are pulled from the top of the project queue.

6. Project team members focus on one or two projects to minimize multitasking.

7. Project team members are assigned no more than 70 percent of capacity.

8. Management focuses on project priorities in the queue.

9. Project details are reported to management on an exception basis.

See sidebar 4.3 for a case study on a resource-constrained company.

Sidebar 4.3. Case Study: How Prioritization Valves Helped a Resource-Constrained Company

A company we worked with had extensive problems getting products launched on time. Most projects were at least 15 to 24 months late. We analyzed the situation and found that, on average, project engineers were spread across five projects. Also, project priorities changed frequently, creating chaos for the project teams, as the engineers had to keep adapting to the new top priority. They couldn't focus on completing work on any specific project.

Our analysis showed the number of active projects was unsustainable with the available resources. We implemented a system to evaluate and optimize the number of projects based on available resources. We worked with the management team and found 35 active projects. This was a wake-up call for management, and they made some significant changes to address these issues. By determining whether to proceed with, place on hold, or stop each project, they decreased the number of active projects from 35 to 9. They also began assigning fewer projects per team member (dropping from five to two). Management was committed to not increasing the number of projects until they added staff to support future growth and contracts.

This initiative made a huge difference. The company began delivering projects on time for the first time in years. The changes also eliminated a lot of the daily firefighting and tension in the enterprise.

One of the most challenging behaviors for the management team to change was adding projects into the pipeline without available resources. Now they add projects into the prioritization valve and actively manage the queue in these ways:

- Adding resources (internal and external) based on the desired project throughput

- No longer sending projects into the process until capacity is available—for example, managing the rate at which projects are made active

- No longer preempting projects but instead allowing them to complete before making a new project active

- Moving projects needing a higher priority to the front of the queue

- Prioritizing projects by sorting them into three color-coded classes:

- Orange for ideas that have not been assessed or prioritized

- Yellow for projects selected to proceed but not yet active, which can move up or down in priority

- Green for active projects that do not change in priority

Projects do not enter the green group until resources are available.

We also demonstrated that prioritizing projects and focusing resources to complete fewer high-impact projects in a shorter time was a more effective use of resources, translating into more profit (Figure 4.15).

Figure 4.15: ROI Improved With Focused Resources

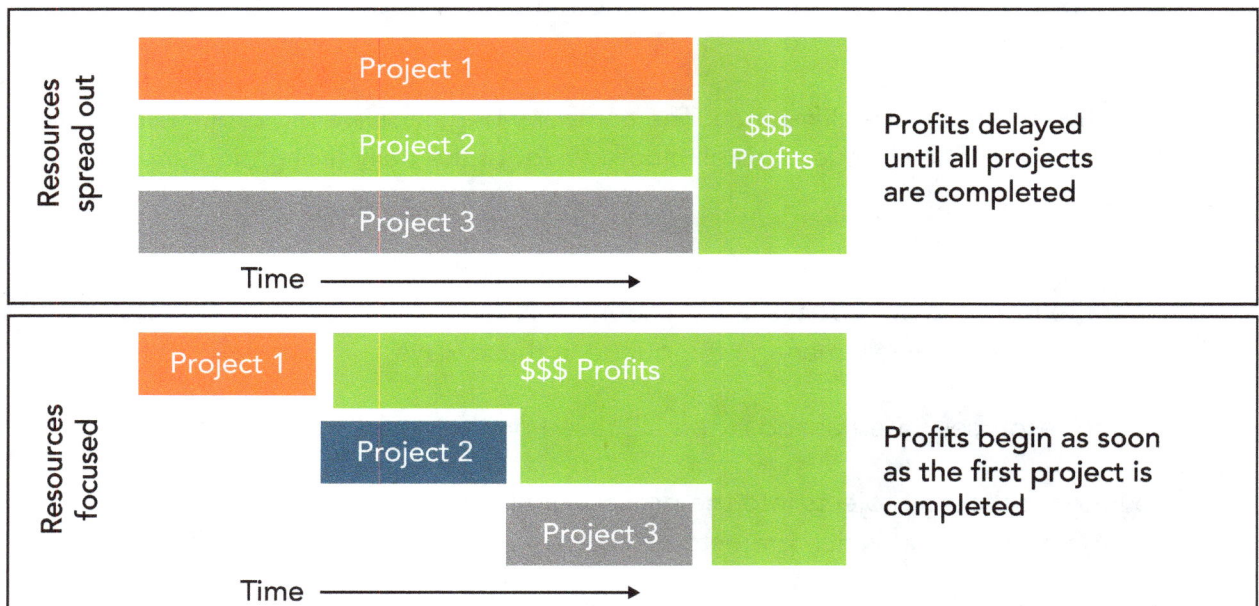

When managing resources, the management committee must determine what resources are constrained, easily acquired and whether they are willing to fill any resource gaps. They also need to consider future resources that are available throughout the entire project. Prioritization starts at PV1, but it is continually updated at other valves throughout the process. To learn how Turba executed PV1, see Sidebar 4.4.

Prioritization starts at PV1, but it is continually updated at other valves throughout the process, where resources need to be managed.

Sidebar 4.4. PERS Is Prioritized

TURBA CORPORATION

As we saw in Sidebar 4.1, at Turba, the PERs product is on the product roadmap and is aligned with Turba's strategy. Also, it was previously vetted by management. Therefore, the idea moved immediately to PV1.

Within PV1, the Turba management committee prioritized PERS as the next product idea in the valve to be released within the ExPD process.

Characteristics of an Effective Management Committee

While the project team gathers ideas, the management committee—which some companies refer to as the Discovery Team, Idea Review Team, or Eureka Team—is responsible for setting and adjusting priorities.

We find that the best management committees are cross-functional. The relevant functions include but are not limited to product management, marketing, design, engineering, operations, a seasoned strategic salesperson, and if you are so fortunate, a Senior Fellow (a broadly experienced engineer or scientist who is typically on a technical track). Depending on the company, the management committee can also include either the president or the general manager.

Besides getting the right people on the committee, success requires timing the committee's involvement in product development and establishing practices to reduce decision-making bias.

Early Involvement

We recommend active management involvement early in the process when idea selection is paramount (Figure 4.16). ExPD has adapted the concept of early management participation from Wheelwright and Clark.[2]

Based on their research, the most successful scenario is management participation that begins at the project's earliest phase and tapers off toward development and launch. Unfortunately, they found low participation early in the process and more participation shortly before launch. We have witnessed this behavior at some companies. It typically results in mayhem for the project team as management makes last-minute changes and contributes to rework as the product enters launch.

We recommend active management involvement early in the process when idea selection is paramount.

Figure 4.16: Management Involvement

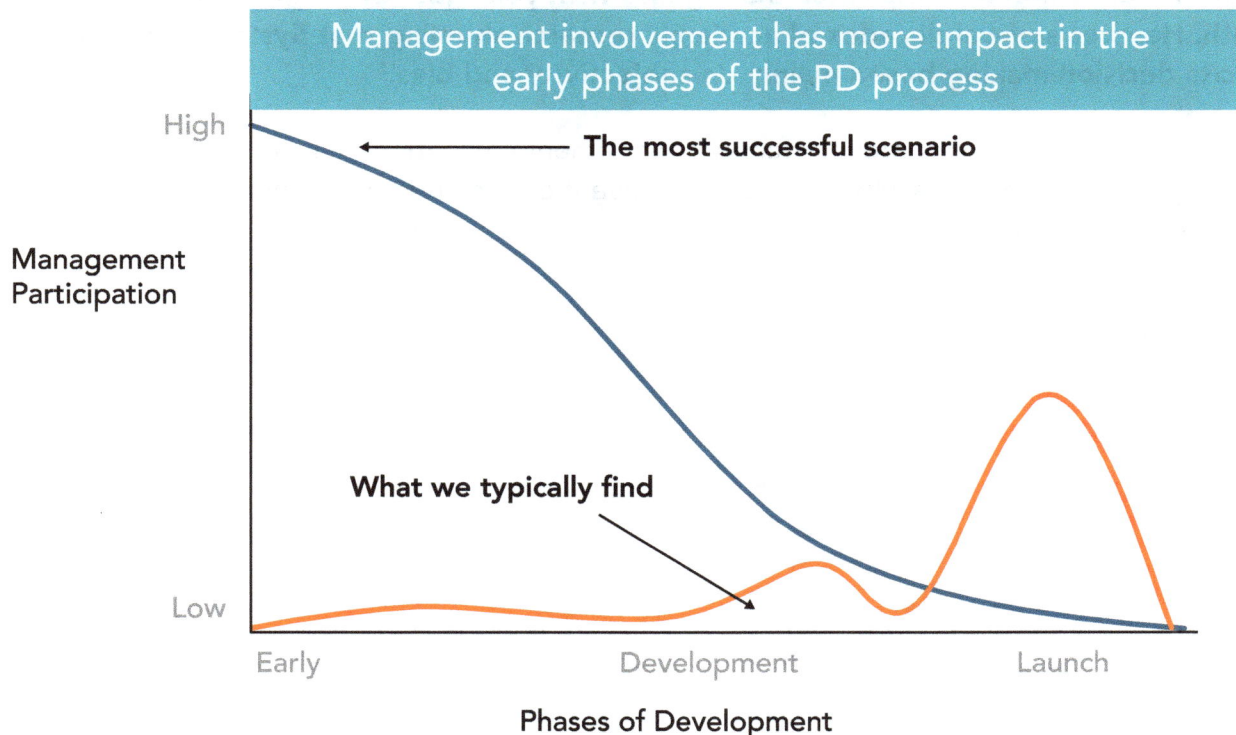

Management involvement has more impact in the early phases of the PD process

High

The most successful scenario

Management Participation

What we typically find

Low

Early   Development   Launch

Phases of Development

Maintaining Objectivity

While implementing product development processes, we also discovered how difficult it was to keep the management committee objective in assessing and selecting ideas. There can be a tendency toward groupthink and agreement with the most senior person in the room. Also, there may be internal biases and decisions based on intuition and emotion, which Kahneman refers to as System 1 Thinking (Sidebar 4.5).

There can be a tendency toward groupthink and agreement with the most senior person in the room.

Sidebar 4.5. Going with the Gut versus Logic-Based Decision-making

In his book *Thinking, Fast and Slow*, Nobel laureate Daniel Kahneman describes a systems model of the cognitive processes involved in decision-making that drive the way we think. System 1 is fast, instinctive, and emotional; System 2 is slower, deliberate, and logical. Kahneman demonstrates that "going with your gut" often goes down the wrong path. He cites several decades of academic research showing that System 1 thinking elicits decision-making based mainly on subjectivity and bias.[3]

In the product development process, management committee members will commonly make gut-based decisions when they do not have the correct data and evidence to make a well-informed decision.

We have outlined six essential practices that assist the reviewer in making the transition from gut-based (System 1) to logic-based decision-making (System 2):

1. **Senior management must be involved early in the ExPD process.** Senior managers have the strategic perspective, budget to allocate resources, portfolio knowledge, and authority to make the right project decisions early in the process.

2. **Reviewers should be cross-functional.** Getting the perspectives of different functional areas seems obvious, but it doesn't always occur. Silos still exist within most enterprises, especially between marketing and engineering. We commonly hear about engineering being left out of the management committee meetings. Perhaps this is due to the culture of marketing and engineering operating at different ends of the product development spectrum. However, sound decisions are based on both commercial acceptance and technological feasibility.

3. **Present content, not a dog and pony show.** Management committees are not where the project teams show off their extraordinary presentation skills. This is a serious investment discussion. It is appropriate for the key project team members to attend the meetings to answer questions, but they should be directed to leave the glitzy presentation behind.

4. **Management committee members need to do their work too.** We understand that these members are commonly senior managers, and they are busy people, but well-informed decisions cannot be based on snap judgment. Reviewing documentation before the meeting can help the committee make a well-informed decision and keep the meeting on track.

5. **Instead of looking for a solution early in the process, look for a need.** Sometimes management committee members request a solution early in the process instead of understanding the unmet customer need. Early in the process, an overly defined solution can lead to missed opportunities for a better solution and later to a misguided product.

6. **Do not expect detailed financials until the product is well defined.** This is where reviewers need to temper their expectations. We worked with a client that expected a detailed financial analysis, including a forecast before the product was defined. This led to a lot of team anxiety and accurate forecasts less than 50 percent of the time—in sum, a lot of unnecessary work provided inaccurate results. We are proponents of estimating financials based on what is known. A detailed analysis is appropriate when the project uncertainty is being resolved and the product is more defined.

The management committee meeting is a pivotal forum in an enterprise, especially in fast-moving markets. A significant cause of waste and delay is premature decisions based on incomplete knowledge or shifting information. Committee members should detach themselves from the emotion, intuition, and gut feelings associated with the idea and instead debate the pros and cons based on sound data and objective criteria. ExPD avoids System 1 decision-making by consciously assessing assumptions and acquiring relevant data and evidence through experimentation and research.

We created the Idea Screening Package to combat groupthink, biases, and gut decisions to keep decision-making as objective as possible. The package includes a description of the project team's idea (prepared by the project team), criteria, and a corresponding scoring model.

We created the Idea Screening Package to combat groupthink, biases, and gut decisions to keep decision-making as objective as possible.

The management committee is responsible for assessing, scoring, and prioritizing the ideas independently before the meeting. Their decisions are then given to an objective party, such as the process owner, responsible for tabulating scores and ranking the ideas for discussion based on the most significant variances.

The ideas that have the most significant variance are first on the agenda for discussion. If the idea is scored the same across the board and there is slight variance, then there is no need for a discussion, and the ideas will be handled accordingly (continue, adapt, hold, or cancel).

If the idea is on the product roadmap, there should not be much discussion or variance in the scores unless market or technology conditions change. If the idea is not on the roadmap, more due diligence and discussion are needed. The upfront work required for the Idea Screening Package keeps ideas objective and, as a bonus, also leads to improved efficiency during the meeting. For an example of how it has worked in practice, see Sidebar 4.6.

Sidebar 4.6. The Idea Screening Package at a Large Conglomerate

The Idea Screening Package worked particularly well for a large conglomerate that we worked with. A senior vice president (SVP) at the company frequently brought to the committee product ideas that family members and friends had asked him to introduce. Members of the committee were afraid to offend the SVP, so they agreed to develop these ideas. Unfortunately, most of the ideas were poor, lacking any commercial or technical merit.

When we ran our initial meeting using the Idea Screening Package, the ideas from the SVP's family and friends were immediately disqualified. Subordinates in the room were relieved that they could decide on a product idea without offending or disagreeing with the SVP since the scores were anonymous. Even more surprising was the relief the SVP felt when telling his friends and family that the management committee had decided not to proceed with their product idea.

In Chapter 5, "Investigate," we will introduce you to some of the most important tasks within ExPD, which include identifying, evaluating, and prioritizing the most impactful uncertainties and risks associated with product ideas.

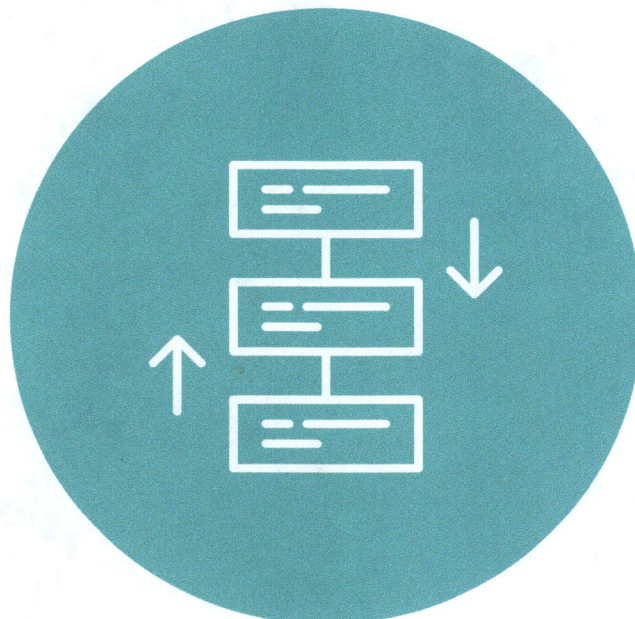

Key Chapter Points

1. Most companies have some variation of the ideation process. What is unique with ExPD is the integrated strategic framework via the roadmaps. Ideas are gathered and enter the pipeline either via the roadmap or as a new opportunity.

2. Ideation is the critical bridge between strategy and development. It includes ensuring that enough of the right ideas are ready to enter the pipeline and requires a robust strategy-driven ideation process.

3. Product roadmaps are not a strategy; they're a graphical representation of the sequence and timing of initiatives or projects. They are very useful in aligning the organization.

4. Many practitioners race to build roadmaps without the necessary product strategies to back them up.

5. In a highly dynamic environment, roadmaps should be broadly directional with relatively short time horizons.

6. A frame is a statement that describes the problem, opportunity, or challenge and delineates the boundaries of an appropriate idea.

7. Looking too close to your existing markets may just turn up the "usual suspect" ideas that your competitors are already exploiting or have already been eliminated for good reasons.

8. Product ideas are captured and filtered in the Idea Library—called a library instead of a repository to express that the team is actively checking ideas in and out.

9. Repeatedly hearing the same ideas may be a symptom of not having an Idea Library to capture and store ideas that enter the process.

10. Some management committee members want to have all the answers about an initial idea, such as its ROI or profit potential, upfront. The management committee must be patient and wait until more is known about the product idea.

11. **A prioritization valve is used to control idea flow based on available resources.**

12. **Active management involvement should begin early in the process when idea selection is paramount.**

13. **In the management committee, there can be a tendency toward groupthink and agreement with the most senior person in the room.**

14. **The Idea Screening Package combats groupthink, biases, and gut decisions to keep decision-making as objective as possible.**

Notes

1. Robert Phaal, Clare Farrukh, and David Probert, "Customizing Roadmapping," *Research Technology Management* (March–April 2004).

2. Steven C. Wheelwright and Kim B. Clark, Revolutionizing Product Development: *Quantum Leaps in Speed, Efficiency and Quality* (New York: Free Press/Simons & Schuster, 1992).

3. Daniel Kahneman, *Thinking, Fast and Slow* (New York: Farrar, Straus and Giroux, 2011).

Segment 3
Explore & Create

In the Explore & Create segment of the ExPD process, **Explore** captures the concept of reducing risk through learning, while **Create** captures developing or transforming an idea into a final product.

Explore & Create consists of three major activities: **Investigate** (Chapter 5), **Plan** (Chapter 6), and **Resolve** (Chapter 7).

During **Investigate,** the project team follows two tracks. Track A identifies, evaluates, and prioritizes the uncertainties and risks that can adversely affect a project. Track B identifies the Idea Maturity Model (IMM) activities, needed to advance the definition, design, and development of the new product.

Upon completion of **Investigate,** the cross-functional team has determined the maturity of the idea, and they have prioritized the most critical assumptions, including any dealbreakers. If the management team decides to proceed with the product idea, a project team is assigned to start the planning process.

Plan entails creating the Product Charter and the Resolve Development Plan (RDP). These two planning documents help the project team to adapt. While the Product Charter provides the project's long-term vision, the RDP provides short-term focus: the top uncertainties to resolve and the activities required to advance the maturity of the product idea.

During **Resolve,** the project team works on resolving the most critical product uncertainties and risks. The Resolve Loop, which includes four steps (Design, Build, Execute, and Learn & Adapt), enables the team to resolve product uncertainty.

Chapter 5
Investigate

Chapter 5 Contents

What to Expect

The input to Investigate is a **chosen idea** (Figure 5.1), which can be a prioritized idea from Prioritization Valve 1 (PV1) or any product idea you want to evaluate. The idea can either be serendipitous or on a product roadmap. Serendipitous ideas often have more uncertainty and require more due diligence. Regardless of the nature of the idea, all ideas proceed through Investigate.

Figure 5.1: Investigate Process

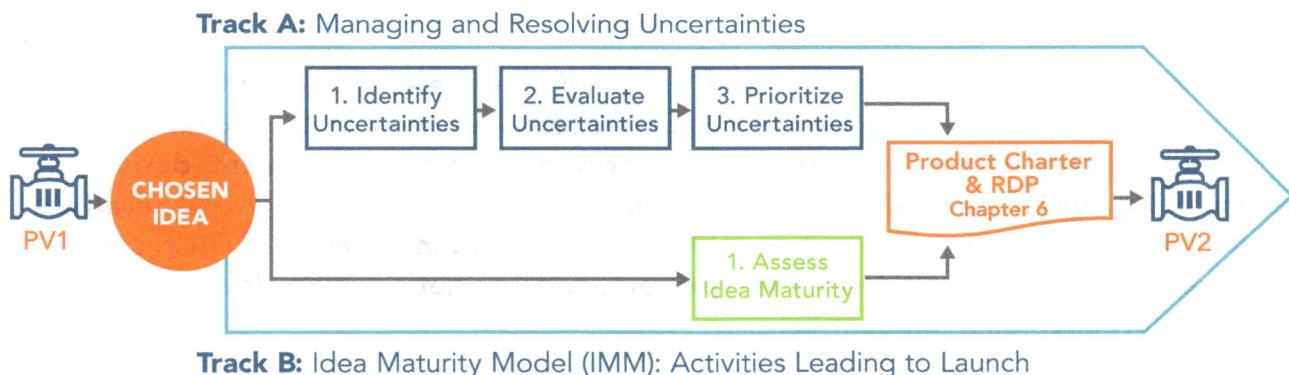

Track A: Managing and Resolving Uncertainties

Track B: Idea Maturity Model (IMM): Activities Leading to Launch

In this chapter, we describe the two major tracks within the Investigate process. In Track A, the most critical product idea uncertainties and risks are identified, evaluated, and prioritized. In Track B, the maturity of an idea is identified within the Idea Maturity Model (IMM), providing guidance on the activities that need to occur to advance the definition, design, and development of the new product.

Track A and Track B inform two documents: Product Charter and Resolve Development Plan (RDP), both of which will be covered in Chapter 6, "Plan."

Getting Started

We recommend that the team prepare for an Investigate session by taking the following three steps:

1. Select a product idea.

The input to the Investigate process, called the chosen idea in Figure 5.1, could be a prioritized idea from PV1, as described in Chapter 4, "Idea Management System." Or it could be any product idea that you want to evaluate.

2. Assign a cross-functional team.

A cross-functional team must be assigned to identify, evaluate, and prioritize the most critical uncertainties. Depending on your organization, we highly recommend that the team members include product management, project management, design, engineering, quality, sales, finance, marketing, regulatory, supply chain, and operations representatives. The benefits are two-fold, the cross-functional team is aware of the proposed product idea and doesn't feel left out, and you get a robust view of the proposed product idea. (Sidebar 5.1).

3. Follow the ExPD methodology.

As mentioned earlier, Investigate consists of two major tracks. Within Track A, our methodology leads the team through identifying, evaluating, and prioritizing the most critical product uncertainties and risks. Track B involves identifying the maturity of the idea within the Idea Maturity Model (IMM). The IMM also guides the product team through the required product development activities.

The methodology guides a thought process without providing a comprehensive checklist of every possible scenario. We have found that this session typically takes up to two hours for a moderately risky product idea and up to four hours for a high risk new-to-the-company product idea.

Sidebar 5.1. ExPD Pilot Findings: The Value of Cross-Functional Teamwork

During an ExPD pilot, we found that the product management team attempted to identify, evaluate, and prioritize product uncertainties independently of other functions. After reviewing their findings, the VP of engineering requested that they go back and involve the cross-functional team to get a more robust view, since there were major gaps in their understanding of the product.

The other members in the second round included engineering, sales, supply chain, and operations/manufacturing. Not surprisingly, the quality and robustness of the uncertainties and risks were superior with cross-functional participation. This exercise provided the team members with a better understanding of the product across different perspectives, enabling them to tackle the product idea holistically.

We also recommend that you include a senior member of staff, sometimes referred to as a Senior Fellow or Member of Technical Staff, to act as a contrarian or subject-matter expert for deep industry experience. If you enter a new category or market, consider bringing in outside experts in this area, like consultants or lead users. For riskier projects—those involving many types of uncertainties to deal with—you may find that an experienced facilitator is very helpful in managing the discussion.

Track A: Managing and Resolving Product Uncertainties

To start the Investigate session, the team identifies, evaluates, and prioritizes the most critical product uncertainties and risks (Figure 5.2).

Figure 5.2: Investigate Process

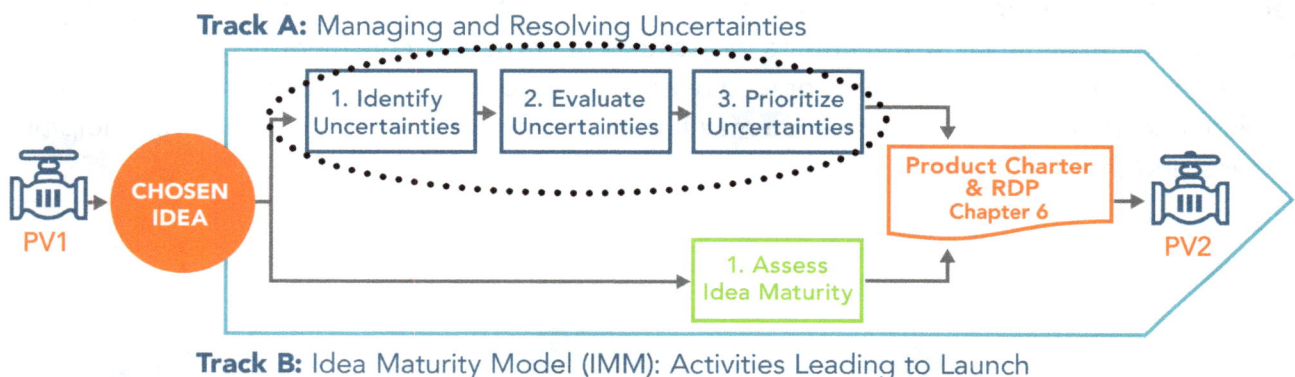

Track A: Managing and Resolving Uncertainties

1. Identify Uncertainties → 2. Evaluate Uncertainties → 3. Prioritize Uncertainties

CHOSEN IDEA — PV1 — Product Charter & RDP Chapter 6 — PV2

1. Assess Idea Maturity

Track B: Idea Maturity Model (IMM): Activities Leading to Launch

Step 1: Identify Uncertainties

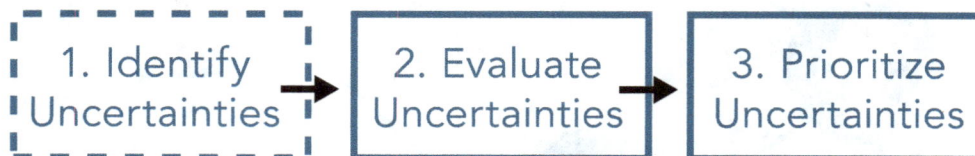

1. Identify Uncertainties → 2. Evaluate Uncertainties → 3. Prioritize Uncertainties

Product developers make investment decisions about product ideas without knowing whether the resulting product will be commercially and technically successful. This uncertainty creates risk. Finding ways to reduce uncertainty is essential to the product idea's ultimate success. The first step is for the team to identify these uncertainties.

Uncertainty in product development arises for many reasons. Here are some signs to look for when identifying product uncertainties:

- Lack of information

- Lack of experience (or being new to something)

- Decisions that have not yet been made

- Outcomes that cannot be known or controlled by the product developers

- The presence of an assumption

Our **Product Risk Framework® (PRF)** tool leads the team through a process to identify, evaluate, and prioritize uncertainties and risks in the four most precarious areas within product development: External Factors, Product/Technical Feasibility, Business Configuration, and Commercial Feasibility (Figure 5.3).

Within the PRF, these four areas, which we call pods support the ExPD methodology. The pods can be adapted for your organization; for example, medical device companies may want to add pods for regulatory and quality uncertainties. Refer to Appendix 5A for more details.

Product-Market Fit, how well the product and market are aligned, is driven by external, commercial and product/technical conditions. Business Fit, how well the product fits the capabilities and strategy of the organization, is driven by commercial, product/technical and business configuration considerations.

Figure 5.3: Product Development Risk Pods

We use the term Business Configuration™ within the PRF tool to refer to the existing resources, assets, activities, processes, technologies, capabilities, and choices that impact, constrain, or enhance the company's ability to develop and support the product going forward. Although the Business Configuration includes the same components as the business model, we extended it to be broader in scope within the PRF. To learn more about the PRF tool, refer to Sidebar 5.2.

Sidebar 5.2: Benefits of the Product Risk Framework (PRF) Tool

The cross-functional team can use either an Excel spreadsheet or the PRF tool for the ExPD process. The value of using the PRF is sixfold:

1. A framework for identifying uncertainties and risks ensures the project team thoroughly identifies all known or discoverable uncertainties

2. Integrated behavioral economic principles, including lenses, nudges, prompts, and reference class forecasting to drive better decision-making

3. Scientific/factual/objective evidence to reduce gut feelings

4. Transparency for the team and increased project visibility and accountability

5. A multidimensional evaluation system identifies uncertainty, impact and how much the uncertainty can be resolved. Uncertainty reduction can be tracked throughout the project, which we refer to as the Assumptions Tracker™

6. A prioritization system based on multiple dimensions of evaluation determines uncertainties and risks that drive project success

To read a case study on how a project team used the PRF tool, please see Appendix 5.1, "Product Risk Framework Case Study—Smarty."

While the cross-functional team identifies the product uncertainties and risks, we recommend that the team reword them as assumptions for a standard framework. Table 5.1 provides examples of rewording uncertainties and risks as assumptions. The project team may find that some uncertainties and risks translate into several assumptions.

Table 5.1: Rewording Uncertainties and Risks as Assumptions

Original Uncertainty/Risk	Assumption
Uncertainty: We don't know if the market for our product idea is large enough to be worth pursuing.	**Assumption:** The number of potential customers is large enough to be attractive.
	Assumption: We understand the critical customer drivers.
Risk: Our lack of experience with this market may result in missteps and rework as we learn about this new market.	**Assumption:** We have a compelling value proposition.
	Assumption: We understand what customers are willing to pay for this product.

Reword uncertainties and risks into assumptions for a standard framework.

When we worked with a pilot team, they found the following benefits of rewording uncertainties and risks into assumptions:

- Evaluation and tracking were more straightforward

- A common language made comparisons easier

- It was easier to identify what needed to be learned to reduce the uncertainty

- Assumptions transitioned directly into the business case

- It helped the team define the product idea better

- Assumptions have more of a positive spin

Before the product has been well defined, there may be many areas of uncertainty, and the team should keep the analysis at a high level. As the team learns more, the assumptions become more granular. It is common for the team to add assumptions as they learn more. Figure 5.4 depicts a typical project over time. Assumptions are added over time, but at the same time, assumptions are being resolved, as explained in Chapter 7, "Resolve."

Figure 5.4: Number of Assumptions Over Time

NUMBER OF ASSUMPTIONS OVER TIME

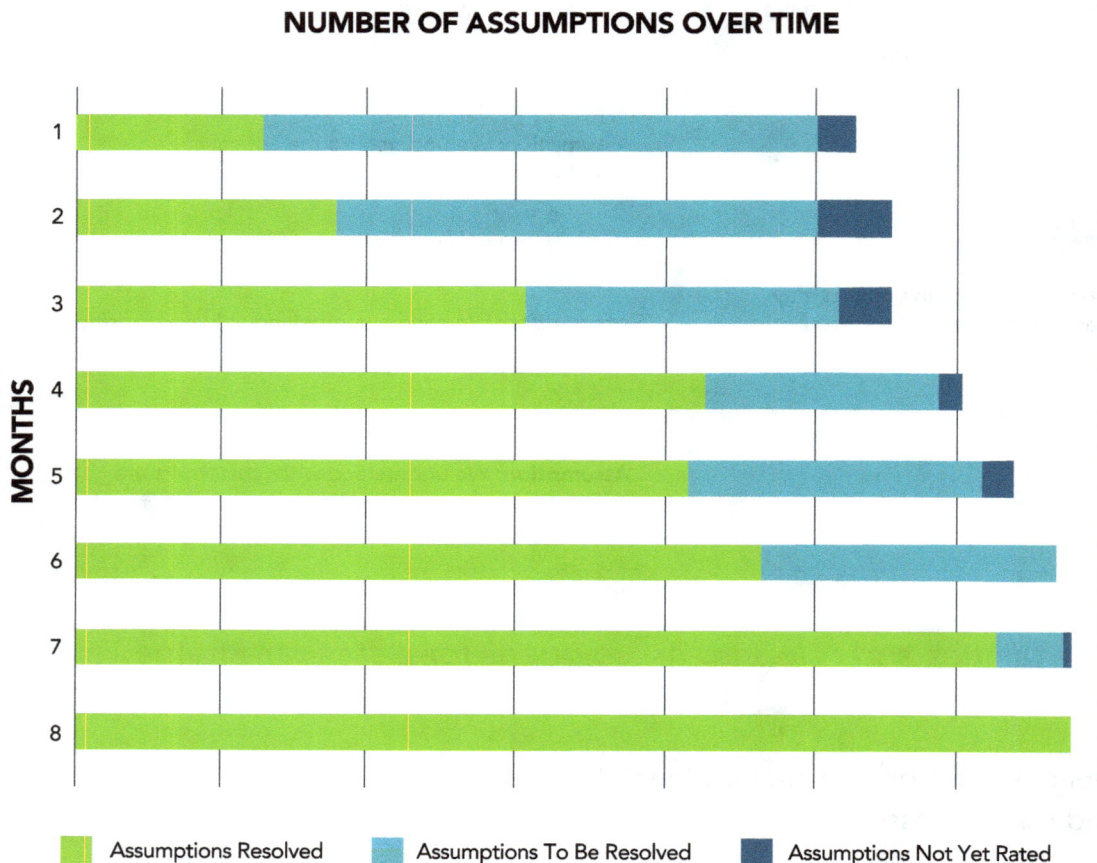

The analysis of assumptions is not static. The team should revisit the assumptions throughout the entire project. Revisiting the assumptions helps the team adapt to any significant internal or external changes that may alter the product's direction. Examples of such changes include unexpected competitor moves, government regulations, and key personnel losses. If these changes are significant and the risk increases, the team must decide whether the company should continue with the project and take on the additional risks.

Revisiting the assumptions helps the team adapt to any significant internal or external changes that may alter the product's direction.

Step 2: Evaluate Uncertainties

1. Identify Uncertainties	→	2. Evaluate Uncertainties	→	3. Prioritize Uncertainties

In Step 2, "Evaluate Uncertainties," the task is to determine each assumption's importance.

Understanding the Risk You're Taking

A common approach is to rate each risk on two dimensions:

1. Probability of occurrence of the undesirable event

2. Magnitude of the impact to the project (dollars lost) when the undesirable event occurs

Teams try to put specific numbers or ranges to probability and impact, but early in the life of a product idea, these estimates are little more than guesses. These estimates become anchors, leading to premature or wrong decisions.

Estimates become anchors, leading to premature or wrong decisions.

There are several dimensions to consider when evaluating and prioritizing risk. The approach to evaluating uncertainty and risk depends on what works best for your team. This section provides two tools for evaluating project risk: Risk Tier and Risk Classification.

Risk Tier. We recommend evaluating assumptions using the Risk Tier:

1. Level of confidence that the assumption is true

2. Impact to the project if the assumption is wrong

Before starting this exercise, we recommend that your team discuss the meaning of "impact" and "confidence" in the context of your project. Historical information from other projects may help define impact and confidence for the new project.

Confidence in an assumption may mean:

a. The team's subjective confidence that the assumption will prove to be true could be described as either highly certain, somewhat certain, somewhat uncertain, or highly uncertain.

b. The level of confidence in the data behind the risk could be categorized as either no information or experience related to the assumption, some anecdotal data supporting the assumption, some secondary data supporting the assumption, or solid experience and data supporting the assumption.

An impact may mean:

a. The schedule to release the product on time is the most critical factor. The company is willing to forgo the budget and scope considerations to achieve the schedule (triple-constraint concept).

b. Some risks may be more critical to an organization—for example, regulatory risks versus schedule risks.

c. Product advantage may be more important when you want to be a leader in the market versus making a higher margin.

d. Time impact—for example, a one-month delay on a four-year project—has less impact than a one-month delay on a six-month project.

e. Cost impact is straightforward relative to the budget. If the project's cost is high—say, running at a daily rate of $100,000—the schedule must be scrutinized for cost and time reduction.

Assumptions can be plotted on the Risk Tier Matrix, as described in Table 5.2. Severe risks (those in the upper right) have the most significant impact on product success. Assumptions rated "caution" are less likely to stop a project but can contribute to rework and delays. For example, a risk may have a significant impact but a low probability of occurrence. Risks rated "acceptable" should be monitored; check them occasionally to ensure that there aren't any major changes that would alter the risk rating.

Bucketing the assumptions allows the team to focus on the assumptions that matter without getting sidetracked by estimating probabilities and dollars.

Table 5.2: Risk Tier

Impact to the Project if the Assumption Proves to Be Wrong	Certainty the Assumption Will Prove to Be True			
	Highly Certain	Somewhat Certain	Somewhat Uncertain	Highly Uncertain
Major Change	Caution	Caution	Severe	Severe
Moderate Change	Caution	Caution	Severe	Severe
Trivial Change	Acceptable	Acceptable	Acceptable	Acceptable

Bucketing the assumptions allows the team to focus on the assumptions that matter without getting sidetracked by estimating probabilities and dollars.

Risk Classification. Next, the team should bucket assumptions into four groups based on risk tier and other factors that affect priority, such as the team's ability to reduce or eliminate the risk. Sometimes the team can do nothing to reduce the risk, so it must decide whether the risk is acceptable. The assumptions can be categorized as a dealbreaker, must address, action recommended, or monitor. This exercise helps the team identify potential dealbreakers for the project. These buckets also help your team understand the project's risk profile and where resolution efforts need to focus. For examples of dealbreakers, see Sidebar 5.3.

Sidebar 5.3. Product Idea Dealbreakers

A dealbreaker is anything that could potentially invalidate a project because the organization cannot take on the risk. Here are three examples of dealbreakers:

1. A medical device company was developing a plastic product. Typically, to get the proper flexibility and durability properties, they would use a plasticizer. But Europe was moving to ban certain types of plasticizers. A dealbreaker assumption was that the team could find a material with the same properties but without the banned plasticizers.

2. An industrial company considered a product that included a sister division's software-based component. The dealbreaker assumption was that the sister division would support customizing the software and integrating the final product's hardware and software.

3. A major dealbreaker for many medical device companies is whether the product will be reimbursed by a payer, such as Medicare or an insurance company. If the product isn't covered, the patient must pay the full price. This could limit sales, since potential customers would be less likely to switch from the existing, covered product. The dealbreaker assumption would be that payers will reimburse for the new product.

Analyzing Risk Graphically

The ratings we have described support three types of valuable analysis:

1. **Evaluating a total product.** Understanding the risk profile of a product helps you understand which project areas (commercial, technical, external, etc.) contain the most uncertainty and risk. That helps the team clarify where to focus its efforts (Figure 5.5).

Figure 5.5: Risk Profile by Pod

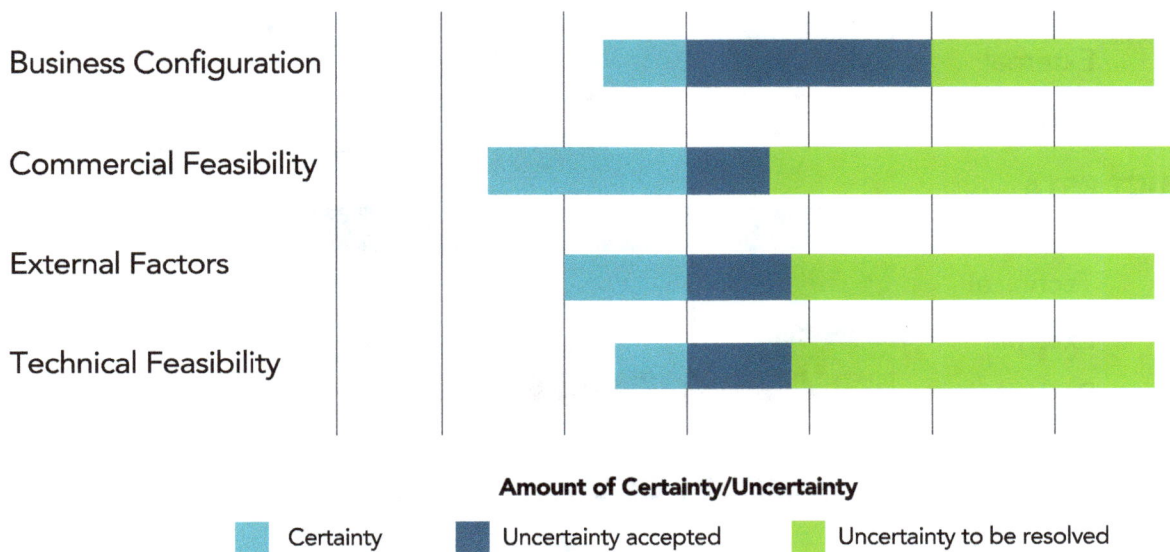

2. **Comparing projects.** Comparing risk profiles helps you understand the level of risk you're taking on with each project. Similarity across current projects may indicate an unbalanced portfolio. For example, the company is being too conservative or too risky. The Product Risk Framework (PRF) incorporates a graphical risk profile (Figure 5.6), so the team can better understand the amount and type of uncertainty and risk they're taking on across a portfolio of projects.

Figure 5.6: Composition of Risks within a Portfolio

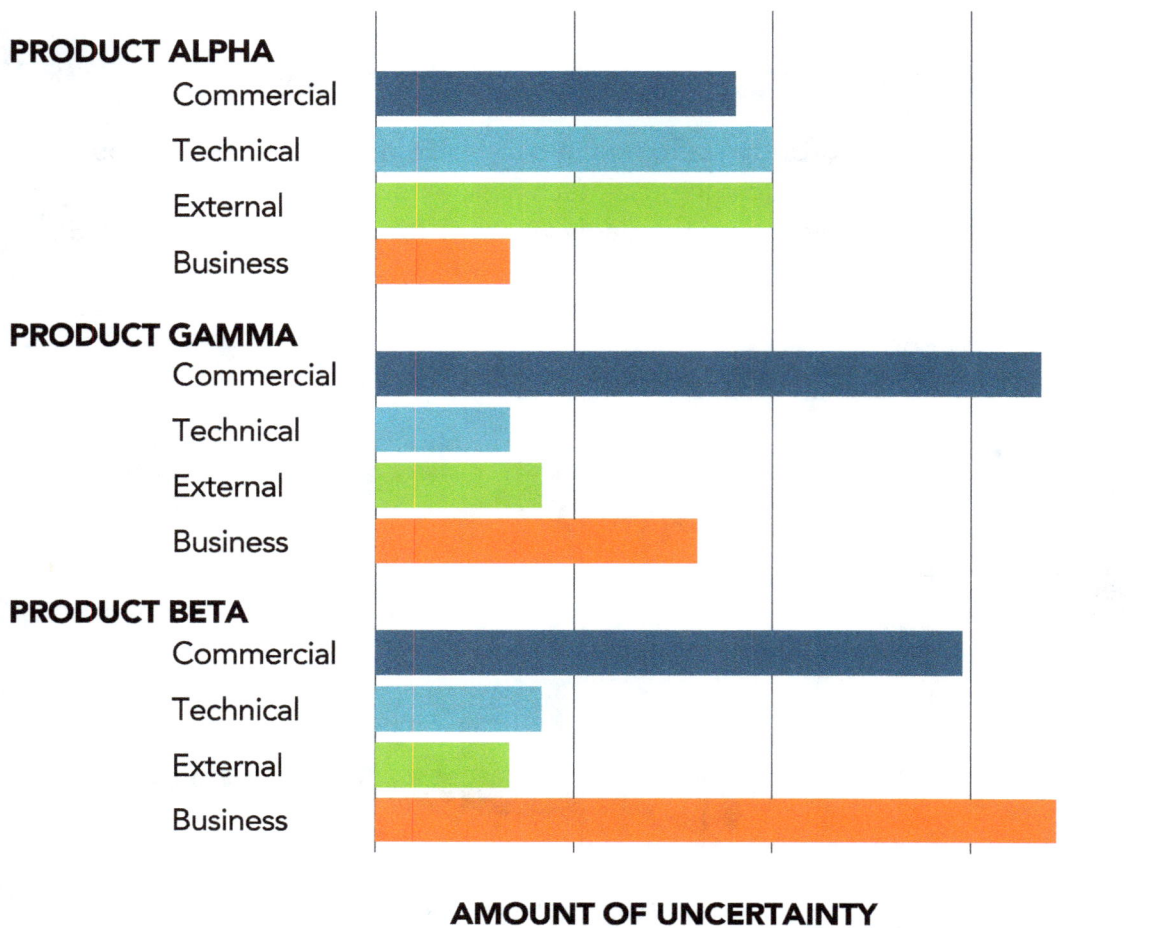

AMOUNT OF UNCERTAINTY

3. **Determining whether a product is a good fit for the company.** Companies should have a clear philosophy on the risks they will adopt with a product idea. Understanding the nature and importance of a product's uncertainty and associated risk is crucial for assessing company fit.

Companies should have a clear philosophy on the risks they will adopt with a product idea.

Step 3: Prioritize Uncertainties

```
┌─────────────┐     ┌─────────────┐     ┌ ─ ─ ─ ─ ─ ─ ─┐
│ 1. Identify │ ──▶ │ 2. Evaluate │ ──▶   3. Prioritize
│Uncertainties│     │Uncertainties│     │ Uncertainties │
└─────────────┘     └─────────────┘     └ ─ ─ ─ ─ ─ ─ ─┘
```

The team needs to negotiate and prioritize the final order of the assumptions. We recommend the following steps:

A: Determine Whether to Proceed

Based on the team's work during identifying and evaluating assumptions, they now need to determine if it makes sense to proceed with the project.

First, are any assumptions classified as a dealbreaker? These are the assumptions that are most likely to invalidate the project and cannot be resolved. Are any assumptions so high in uncertainty and impact that the company cannot accept the risk? Remember, canceling the project quickly can lead to a reallocation of resources to more valuable projects.

Next, does the product idea align with your company's desired product risk profile? These are the areas where the company feels comfortable taking risks, including technical, commercial, and regulatory risks.

Can your team resolve the product uncertainty? There may be so much unresolvable uncertainty with the product idea that management must decide whether to cancel or proceed with the product.

B: Force Rank the Assumptions

If you decide to proceed with the product, prioritize your assumptions. It is essential to have your cross-functional team discuss and negotiate how the assumptions should be resolved. A resolved assumption is one where the team has decided the level of risk is acceptable. During this ranking discussion, compare assumptions against each other to determine the order in which they should be resolved.

As part of this discussion, you should consider the following criteria:

- Quick hits, which you may want to move to the top of the list

- Company's risk attitude (willingness to take on risks)

- Amount of reduction in uncertainty or impact that can be realized

- Cost to resolve the assumption, including the number of resources and time required

- Potential to group some assumptions and resolve them together

- Potential to resolve some assumptions incrementally—for example, by getting more detailed or precise over time

To see how Turba approached the Investigate process, see Sidebar 5.4.

TURBA CORPORATION

Sidebar 5.4. Turba's Investigate Session

Turba's cross-functional product development team—including product management, project management, design, engineering, finance, advanced development, sales, operations, marketing, and a senior fellow met for a two-hour Investigate session for the PERS product. Since this idea was on the product roadmap, it had been vetted and didn't require as much due diligence as a serendipitous idea.

During the Investigate session, the team used the business model developed during the strategy session (see Chapter 3, "Why Strategy Matters for Product Development"). The business model was an excellent tool for the team to use during this session since it focused on how all the system elements fit into a working whole. It provided a framework for thinking through all the levers that can be tweaked to create and deliver the PERS solution. The team focused on the gaps in the business model to identify any assumptions (see dotted lines in Figure 5.7).

Figure 5.7: Turba's PERS Business Model

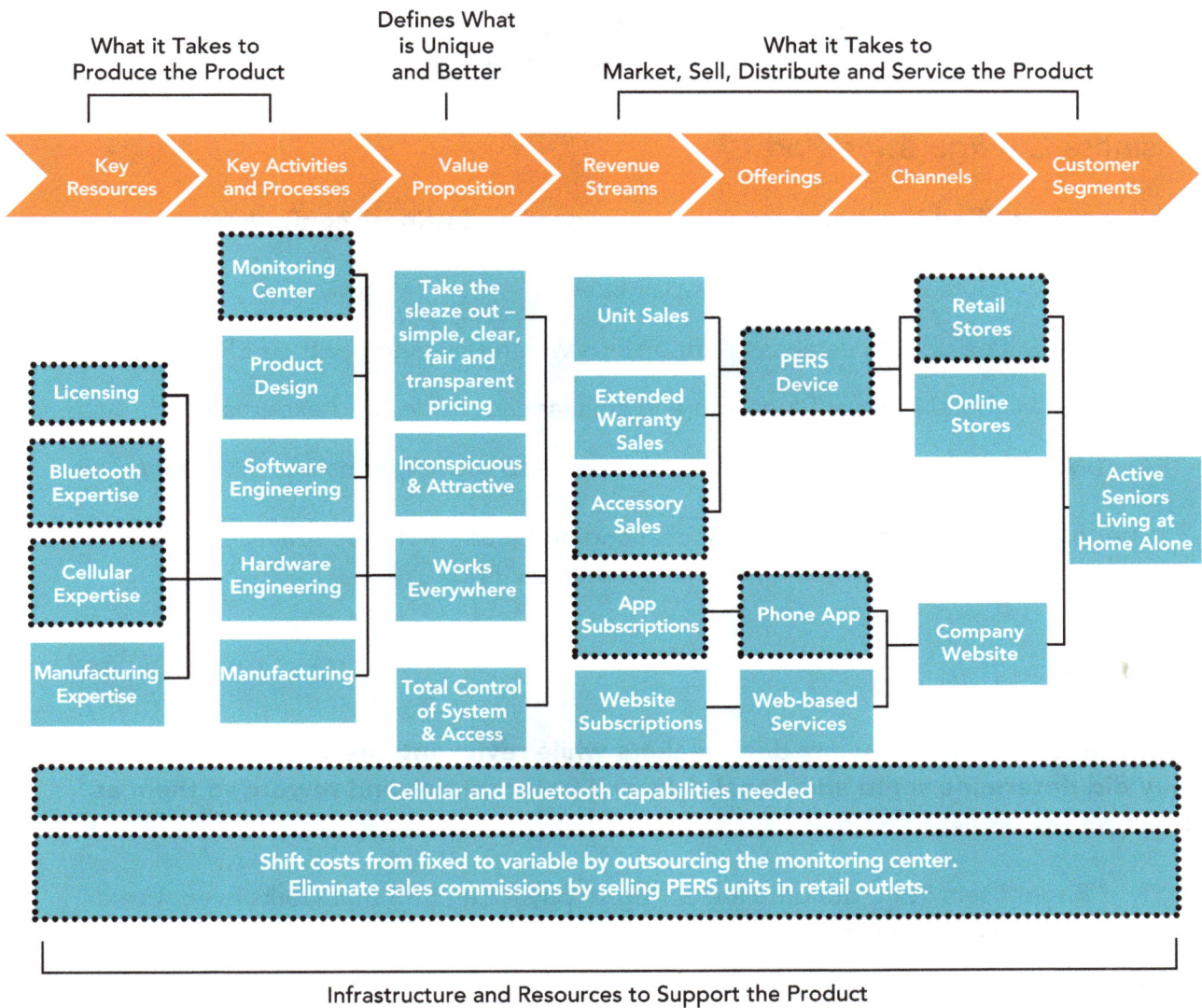

	What it Takes to Produce the Product		Defines What is Unique and Better	What it Takes to Market, Sell, Distribute and Service the Product			
Key Resources	Key Activities and Processes	Value Proposition	Revenue Streams	Offerings	Channels	Customer Segments

- Monitoring Center
- Product Design
- Software Engineering
- Hardware Engineering
- Manufacturing

Key Resources:
- Licensing
- Bluetooth Expertise
- Cellular Expertise
- Manufacturing Expertise

Value Proposition:
- Take the sleaze out – simple, clear, fair and transparent pricing
- Inconspicuous & Attractive
- Works Everywhere
- Total Control of System & Access

Revenue Streams:
- Unit Sales
- Extended Warranty Sales
- Accessory Sales
- App Subscriptions
- Website Subscriptions

Offerings:
- PERS Device
- Phone App
- Web-based Services

Channels:
- Retail Stores
- Online Stores
- Company Website

Customer Segments:
- Active Seniors Living at Home Alone

Cellular and Bluetooth capabilities needed

Shift costs from fixed to variable by outsourcing the monitoring center. Eliminate sales commissions by selling PERS units in retail outlets.

Infrastructure and Resources to Support the Product

Also, the team used the Product Risk Framework (PRF) tool to identify any additional uncertainties and risks. The team then evaluated and reworded the uncertainties and risks as assumptions. The team started with the Business Configuration Pod.

Business Configuration Pod

The team identified the major assumptions regarding business configuration:

1. "We will outsource monitoring capabilities."

2. "We will be able to establish partnerships with retailers to sell the PERS device."

3. "We will build internal Bluetooth and cellular capabilities and expertise."

4. "We will realize additional revenue through app subscriptions and accessory sales."

The team determined that the most challenging assumption was **selling the PERS through retail outlets;** they then continued to the Commercial Pod.

Commercial Pod

The team did not identify any dealbreakers while reviewing the Commercial Pod, but they did determine some additional risks and uncertainties and reworded them as assumptions, as listed below:

1. "Target users living at home are willing to pay for a PERS that works everywhere, including outside the home."

2. "The market is big enough and growing fast enough to support the PERS product."

3. "The product value proposition is compelling to the customer."

4. "PERS pricing is attractive to the customer while still making a profit."

5. "Consumers will purchase PERS units through retailers."

6. "Turba will be recognized as a trusted, reliable brand in retail channels."

7. "Technology has become mainstream, so consumers are willing to adopt."

The team ranked **willingness to pay for a PERS that works everywhere** as the top uncertainty within the Commercial Pod. The team then continued to the External Factors Pod.

External Factors Pod

The team identified the following external uncertainties of the PERS product:

1. "The PERS product pricing is not sensitive to either an upturn or downturn in the economy."

2. "The PERS product aligns with the significant demographic shift of an aging population."

3. "The latest style and technology trends align with the active Baby Boomer generation."

4. "The PERS product's profitability will not be affected if sold in different countries or regions abroad."

5. "The product will obtain UL, CSA, and CE marks."

6. "The PERS product will use commonly available components to avoid supply chain disruption."

7. "Competitive activity will not effect the PERS product at launch."

The team felt unsure about future competition since there were signs of disruption in the PERS category. They decided to monitor any potential competitive activity because future outcomes were unknown. The team then tackled the Product/Technical Pod.

Product/Technical Pod

The cross-functional team didn't identify any technology dealbreakers, but that might change once they start developing the product. The following technical assumptions were identified:

1. "Bluetooth technology will reliably enable PERS to work anywhere a smartphone works."

2. "There are no disruptive technologies on the horizon."

3. "We will be able to build a new platform for the PERS category."

4. "Our fast-follower approach to innovation aligns with leveraging mature technologies."

5. "IP is not an issue with the PERS product since we expect to license the technology."

6. "We will manufacture this product without any significant changes to our manufacturing processes."

7. "We will design an inconspicuous and attractive PERS with a simple user interface."

The team ranked **Bluetooth technology will reliably enable PERS to work anywhere a smartphone works** as one of the most critical uncertainties within the Product/Technical Pod.

Although the team prioritized all the assumptions within each pod, the following two assumptions were identified as the most critical product uncertainties to resolve:

1. **Target users living at home are willing to pay for a PERS that works everywhere, including outside the home.**

2. **Bluetooth technology will reliably enable PERS to work anywhere a smartphone works.**

During the session, the project manager recorded all the assumptions within the PRF tool. The Assumptions Tracker embedded within the PRF will enable the project team to keep tabs on the resolution and addition of assumptions during the project.

The team members took a 10-minute break before starting the Idea Maturity Model (IMM) exercise (Sidebar 5.7).

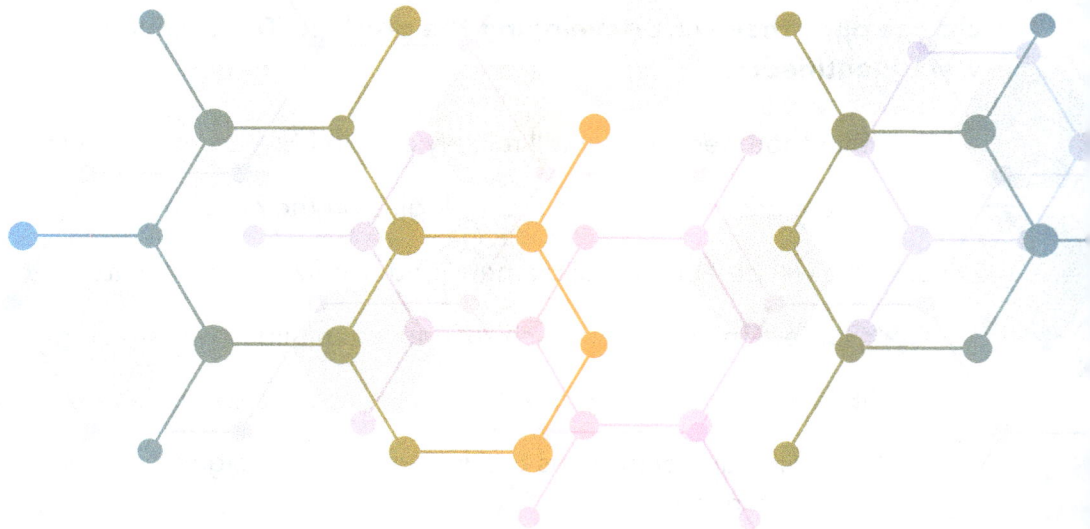

Track B: Product Development Activities

While resolving product uncertainty is very important to product success, you also need to launch a market-ready product. ExPD recognizes that product development is more than resolving uncertainties. Once a project team understands the level of product uncertainty, they can begin planning the product development activities. This is where the Idea Maturity Model comes into use (Figure 5.8).

Figure 5.8: Assess Idea Maturity

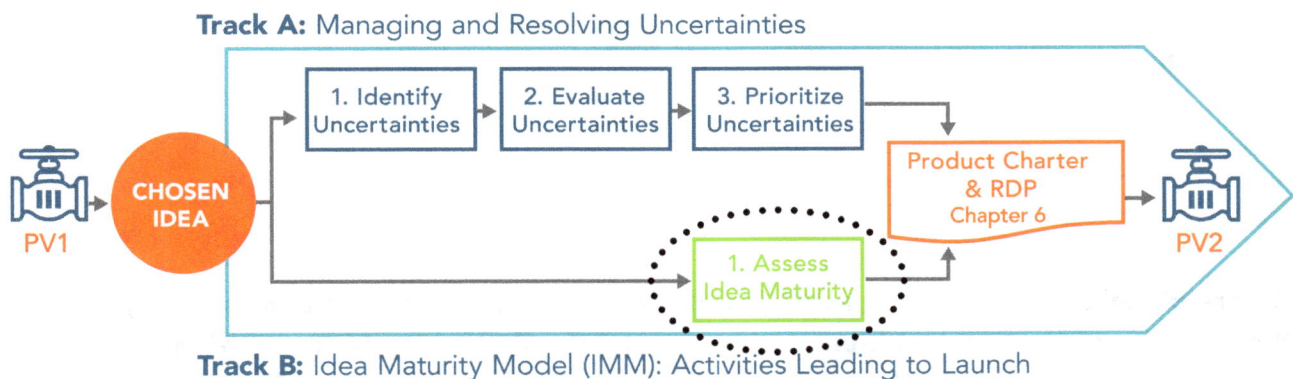

Track A: Managing and Resolving Uncertainties

CHOSEN IDEA

PV1

1. Identify Uncertainties

2. Evaluate Uncertainties

3. Prioritize Uncertainties

Product Charter & RDP
Chapter 6

PV2

1. Assess Idea Maturity

Track B: Idea Maturity Model (IMM): Activities Leading to Launch

ExPD recognizes that product development is more than resolving uncertainties.

Product development includes many activities that lead to a developed product, such as the following:

- Technical development

- Verification and validation

- Production setup, including documents, drawings, ERP, and process validation

- Supply chain, including vendor qualifications

- Marketing preparation, forecasting, and sales training

- Customer service preparations, including field service and internal call centers

- Regulatory and certification filings

- Launch activities

Assess Idea Maturity

Before planning development activities, the team must understand their starting point: the maturity of the product idea. We've created the ExPD Idea Maturity Model™ (IMM) to help teams understand the maturity of the idea and what is needed to evolve the project toward a final product (Table 5.3). The IMM was adapted from the Crawford and Di Benedetto Concept Life Cycle.[1]

Table 5.3: ExPD Idea Maturity Model™

Choose your starting point:

Increase product maturity and reduce risk

Maturity Level	Description	Key Activities & Tools	Launch Activities
Opportunity	Company skill or resource to be leveraged; a customer problem to solve; a competitive threat to be responded to, etc.	Where will we invest our resources?	None
Idea	Initial thought on how to address the opportunity	Brainstorming, customer solution, problem solving, competitive analysis, etc.	None
Stated Concept	Potential solution(s) to address the opportunity/problem to be solved, plus a clear statement of benefits. Hypothesized value proposition. This is the gateway to research and experimentation	Prototypes, artist renderings, Minimum Viable Product (MVP)	None
Tested Concept	Selected concepts have passed end-user concept testing; statements of benefits are confirmed; value proposition is validated. Potential technologies identified	Execute concept testing, prioritized needs, human factors analysis, and concepts that move forward	Value proposition for market positioning
Targeted Concept	Concept is defined: selected user needs and product requirements to address them, the benefits to be provided, plus any mandatory features, claims, functionality. Feasibility of technology proven beyond a reasonable doubt	Translate customer needs into technical requirements, preliminary Design for Manufacturing (DFM), high-level architecture, complete Product Viability Analysis	Identify evidence to support market claim, preliminary forecast
Designed Concept	A tentative physical product is created, including features and benefits. Final product design transitioning to manufacturing	Alpha and beta testing, Verification and Validation (V&V), Bill of Material (BOM), design documentation	Agency approvals, launch plan, market plan, operations plan (factory, distribution points) SKUs generated, labels, package
Manufacturing Concept	The full manufacturing process is complete. A supply of the new product produced in quantity from a pilot production line	Testing manufacturing process, tooling, quality plan, V&V	Finalize marketing materials, sales, customer service training, first production run, post-launch plan, first article
Launch	Full production	Product roll-out	Kickoff marketing and operational plan

Let's look at each of the elements of the IMM (Table 5.3):

Maturity Level: This refers to different levels of idea maturity, from opportunity through launch.

Loops: The looping arrows indicate a potential iteration between the maturity levels.

Description: This is a high-level description of what occurs within each level of maturity, expressed in terms of that level's end state.

Key Activities & Tools: These are suggested activities and tools that can be used at different levels of maturity.

Launch Activities: These recommended activities help to ensure an on-time launch. Launch activities are typically overlooked until too late in the process, leading to significant project delays.

You may find it helpful to modify the IMM to align with your company's internal practices and activities, as we did with each of our pilot companies. Also, the IMM can be modified for your organization and industry type. For example, we added a column of design control deliverables for a medical device customer. We also recommend that you tie the maturity levels to your project plans.

The IMM can be modified for your organization and industry type.

The IMM delivers three significant benefits:

1. Determines the project's starting point.

2. Provides structure for the project.

3. Identifies additional uncertainties.

1. Determines the Project's Starting Point

During our pilots, we found that it was important for the teams to initially describe their idea at a very high level and discuss where it was located on the IMM. The IMM helped teams orient themselves to the product's level of maturity, and it gave them a starting point for the project, as described in Sidebar 5.5.

Sidebar 5.5. "Where Is This Idea on the IMM?"

A client asked the following question: "There are many pieces of wearable technology available, each monitoring one or two things well. However, consumers need all the data pulled together and analyzed in an actionable way. There is an idea in there somewhere, but it needs a lot of work. Would this be listed as an 'Opportunity' within the Idea Maturity Model?"

We answered, "Yes, that's how we would define it since there's a customer problem to be solved, but it isn't quite an 'Idea' because there isn't a potential solution yet."

Ideas on the product roadmap typically start at the level of an Idea or a Stated Concept. Still, some ideas, especially serendipitous ones, could be very poorly defined and start at the Opportunity level. In that case, the team must flesh out the idea and conduct the necessary due diligence to determine if the project is worth pursuing further.

A project can also start at a greater level of maturity. For example, if a company buys technology or a license, the starting point could be from a Tested Concept to a Launched product.

2. Provides Structure

The IMM is an integral part of the ExPD process since objectives will be set based on the project's maturity level. The IMM guides the team's progress toward a market-ready product. The maturity of the idea evolves over the project's life, and the IMM assists in measuring the team's progress toward Launch.

The IMM guides the team's progress toward a market-ready product.

For example, suppose the idea enters Investigate as a Stated Concept. In that case, the team determines what activities need to be completed to get the project to the next level of maturity, Tested Concept. This results in clear guidelines, timelines, and goals. For the team, these parameters create urgency and the need for action.

For management, these parameters identify the resources and budget available to the team to execute their plan. These agreements formalize the commitments made by both the project team and management to move the project to Tested Concept.

The IMM also provides guidance on bringing in extended team members to contribute to the product deliverables. For example, your manufacturing partners would be brought in to create the Design for Manufacturability (DFM) as the project evolves to a Targeted Concept. Extended team members also contribute to the Product Viability Analysis (Sidebar 5.6). The Product Viability Analysis is used to demonstrate the product's viability before entering big spend at the Designed Concept level.

Sidebar 5.6. Product Viability Analysis

We developed the Product Viability Analysis to help our client teams demonstrate the product's viability prior to entering development activities during the Designed Concept level within the IMM. The Product Viability Analysis is a cross-functional document that requires the following types of information and data:

- Market summary (market interest, user needs, five-year sales forecast, market strategy, etc.)

- Design and development summary (product design summary, testing activities, etc.)

- Operations summary (operations assessment, DFX, product processes, and supply routes, etc.)

- IP, legal, and regulatory summary (intellectual property, legal status, regulatory requirements, strategy, etc.)

- Financial summary (key financial indicators, investment requirements, etc.)

- Uncertainties and risk summary (for each area listed)

Ideally, the team resolves the most critical uncertainties and risks before entering the Designed Concept level, where expensive development activities occur.

3. Identifies Additional Uncertainties

As part of evaluating the idea's maturity, you may also uncover additional assumptions, uncertainties and risks. While the IMM provides a structure for product development, the project team must always be alert to new uncertainties that need to be added, evaluated, and prioritized.

The project team must always be alert to new uncertainties that need to be added, evaluated, and prioritized.

While uncertainty identification and management are introduced in this chapter, they must be continually managed throughout the project since it is an adaptive iterative process. Some activities will need to be added, deleted, or modified in the project schedule, as covered in Chapter 6, "Plan."

To learn how Turba completed its Investigate session for the PERS project, see Sidebar 5.7.

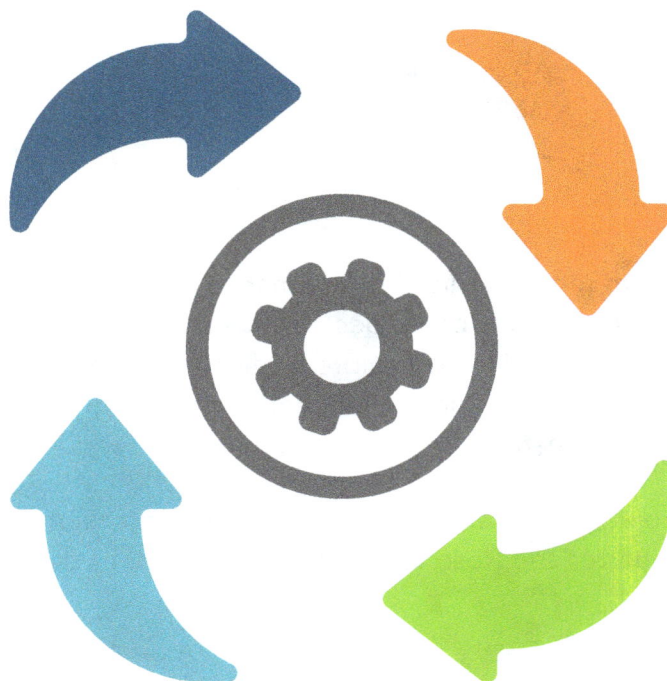

Sidebar 5.7. Turba Wraps Up the Investigate Session

TURBA CORPORATION

The team returned from their 10-minute break and discussed the placement of the PERS product within the Idea Maturity Model (IMM). The next step is to confirm the needs and benefits. They agreed that the product was at the Idea level (Figure 5.9).

Figure 5.9: Maturity Levels of the IMM

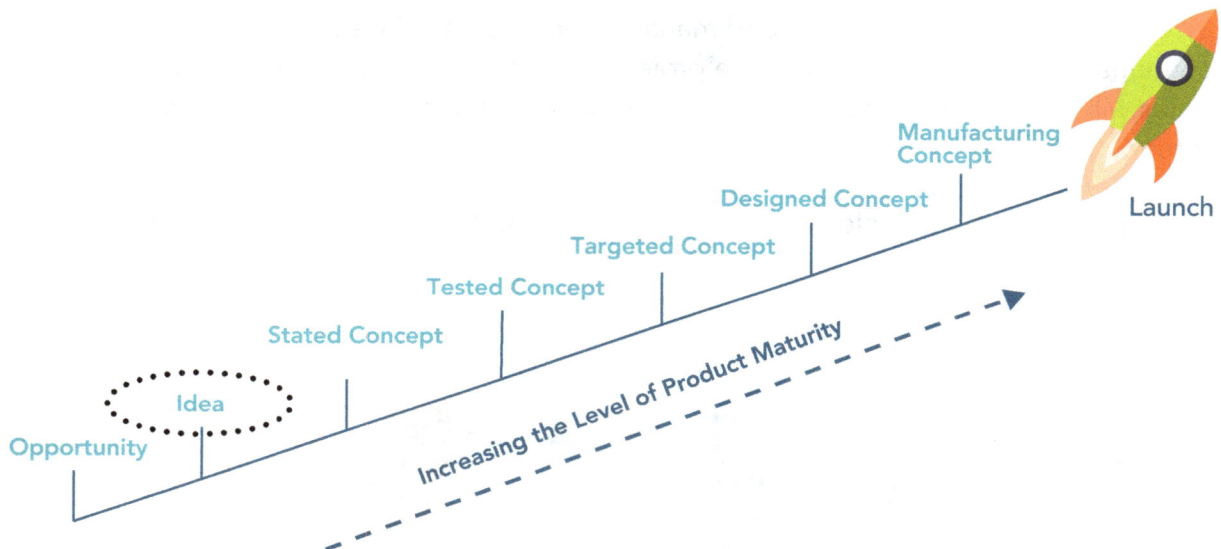

The cross-functional team recommended to the management committee that the PERS product should proceed in the ExPD process because the results seemed encouraging, and they had not identified any significant dealbreakers. They were also confident that they could reach the IMM level of Stated Concept. Management agreed to assign the appropriate resources. The cross-functional team wrapped up the two-hour session.

Next, the assigned project manager will work with the project team members in developing the Product Charter and Resolve Development Plan, as described in the next chapter, "Plan."

To learn more about how another team used the PRF tool refer to appendix 5A.

Key Chapter Points

1. **Investigate runs on two major tracks. In Track A, a cross-functional team identifies, evaluates, and prioritizes the most critical product idea uncertainties and risks.**

2. **In Track B, the team identifies the maturity of the idea within the Idea Maturity Model (IMM), which guides the required product development activities.**

3. **Understanding the risk profile of a product helps you understand which areas (commercial, technical, external, etc.) of the product contain the most uncertainty and risk to determine whether a product idea warrants an investment.**

4. **Four of the most common areas of product development risks are commercial feasibility, external factors, product/technical feasibility, and business configuration.**

5. **We use the term Business Configuration within the ExPD methodology to refer to the resources, assets, activities, processes, technologies, capabilities, and choices that impact, constrain, or enhance the company's ability to develop and support the product going forward.**

6. **While the cross-functional team identifies the product uncertainties and risks, we recommend that the team reword them as assumptions for a standard framework.**

7. **The ExPD methodology guides the team in identifying important assumptions that are commonly overlooked. The team can then develop the means to learn confidently and experiment to reduce uncertainty and risk.**

8. **Revisiting the assumptions helps the team adapt to any significant internal or external changes that may change the product's direction.**

9. When teams try to put specific numbers or ranges to probability and impact early in the life of a product idea, these estimates are little more than guesses, and become anchors, leading to premature or wrong decisions.

10. The highest-priority assumptions have the potential to cancel a project. Addressing these first allows for early termination decisions and redeployment of valuable resources.

11. Bucketing the assumptions allows the team to focus on the assumptions that matter without getting sidetracked by estimating probabilities and dollars.

12. While resolving product uncertainty is very important to product success, you also need to launch a market-ready product. ExPD recognizes that product development is more than resolving uncertainties.

13. The Idea Maturity Model (IMM) is an integral part of the ExPD process. It provides the necessary product development activities to launch the product. The maturity of the idea evolves over the project's life, and the IMM assists in measuring the team's progress toward launch.

Appendix 5A: Product Risk Framework Case Study—Smarty

This case study illustrates a company's use of the Product Risk Framework (PRF) tool that supports the ExPD process. It has been modified to protect our client's intellectual property.

Company: Industrial division of a Fortune 500 company.

Background: A new product concept (nickname: Smarty) would incorporate smart technology into a mature product category. The device is used in freighters, including containers and bulk. The application of smart technology to this device would be new to the category and new to the business unit. Smarty was prompted by a preemptive strike because there was a concern that smart technology in adjacent categories might be incorporated into other products that this customer segment uses.

In freighters (cargo ships), space is limited, so reducing the size of the device was an ongoing need. Another ongoing need was reducing costs. Compared with the current 45-year-old product, Smarty would be much smaller and could fit into a future freighter control room. Additionally, it would reduce operating costs through its failure prevention service (continuous monitoring and predicting).

Smarty Team Session: The cross-functional team representing product management, engineering, operations, supply chain, sales, and marketing attended the Smarty PRF session. During this session, the team evaluated each pod within the PRF, including Commercial Feasibility, External Factors, Product/Technology Feasibility, and Business Configuration. Each pod had six cells with a set of assumptions (Figure 5A.1). The PRF prompted the team to identify, evaluate, and prioritize the most critical and impactful assumptions. Also, the team could flag any constraints that Smarty couldn't meet or any potential dealbreakers.

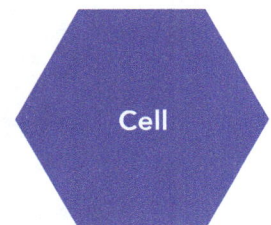

Cell

Figure 5A.1: Product Risk Framework (PRF)

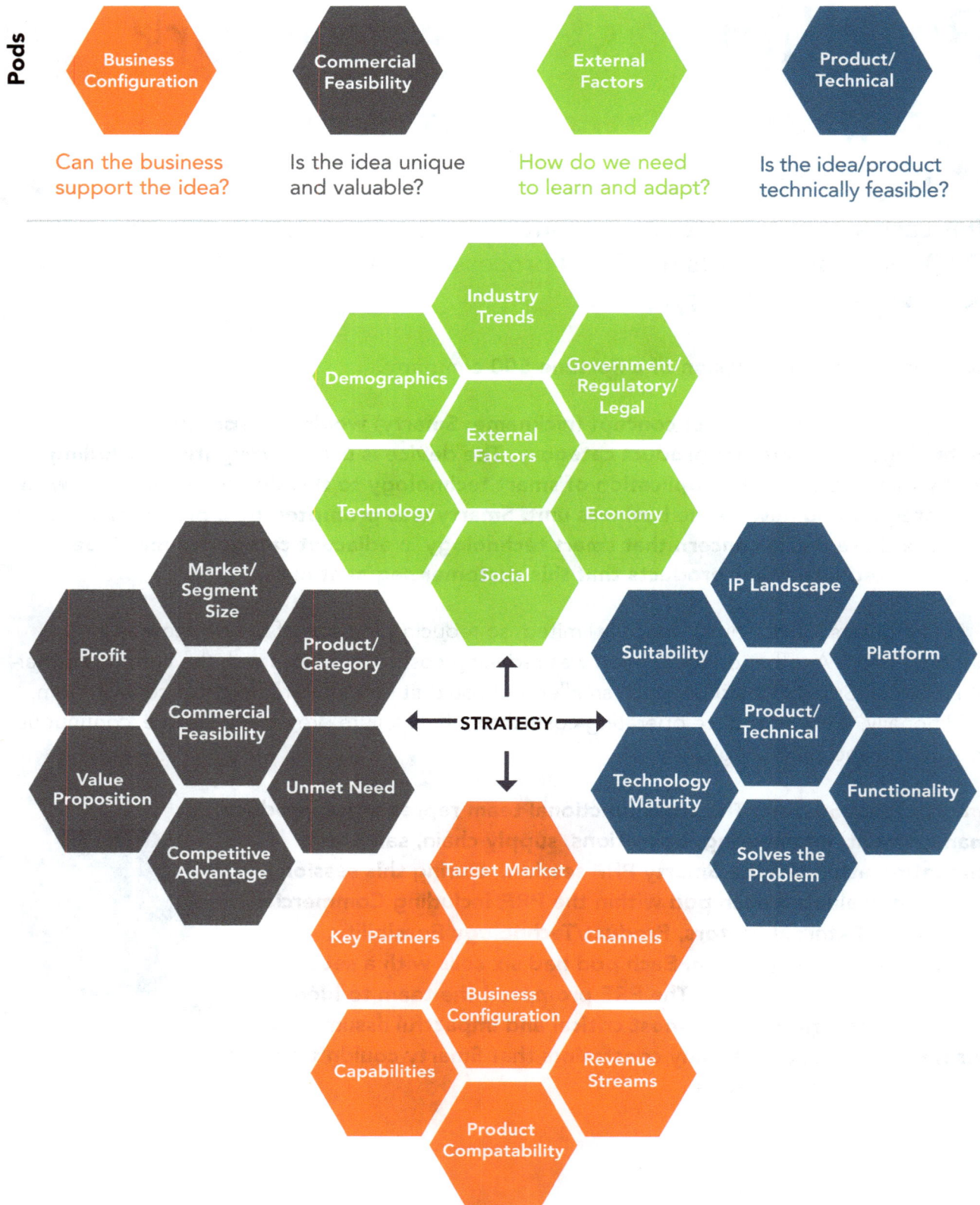

Pods

Business Configuration	Commercial Feasibility	External Factors	Product/ Technical
Can the business support the idea?	Is the idea unique and valuable?	How do we need to learn and adapt?	Is the idea/product technically feasible?

External Factors:
- Industry Trends
- Demographics
- Government/ Regulatory/ Legal
- Technology
- Economy
- Social

Commercial Feasibility:
- Market/ Segment Size
- Product/ Category
- Profit
- Value Proposition
- Unmet Need
- Competitive Advantage

Product/Technical:
- IP Landscape
- Suitability
- Platform
- Technology Maturity
- Functionality
- Solves the Problem

STRATEGY

Business Configuration:
- Target Market
- Key Partners
- Channels
- Capabilities
- Revenue Streams
- Product Compatability

Commercial Feasibility Pod

The PRF tool contains assumptions organized by cell within pod. These assumptions act as prompts or thought starters to help the team evaluate each dimension of the product that may contain risk. The assumptions are not a comprehensive checklist and should be modified for each organization and industry. They should also be updated over time as teams learn.

A product's commercial feasibility refers to whether the product will be differentiated and valuable. In evaluating the cells within the Commercial Feasibility Pod (Figure 5A.2) the team determines whether there are any sizable constraints or any critical uncertainties to be resolved.

The team decided to start evaluating the Commercial Feasibility Pod since they were most comfortable with this aspect of the product.

Figure 5A.2: Cells in the Commercial Feasibility Pod

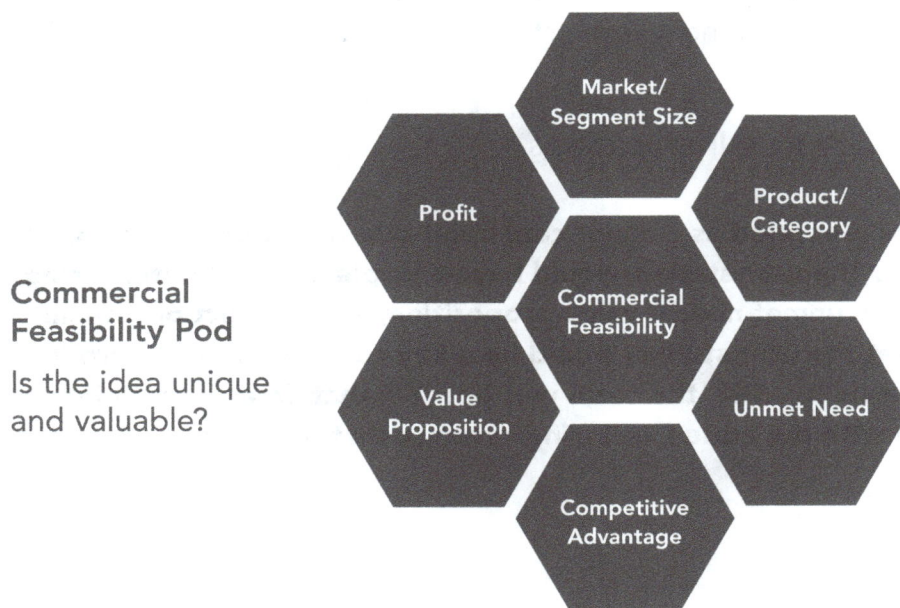

Commercial Feasibility Pod

Is the idea unique and valuable?

Market/Segment Size

Profit

Product/Category

Commercial Feasibility

Value Proposition

Unmet Need

Competitive Advantage

Market/Segment Size. Since the business unit was currently servicing this market, the team thought the market size was large enough. However, market research and user studies were needed to determine how many operators of freighters were willing to buy Smarty.

Product/Category. The business unit was very familiar with this product category. The team was committed to the category, and team members believed they could extend its life cycle with the smart-technology enhancements.

Unmet Need. The team members understood customers' need for a compact size, but they didn't know how other needs evolved. The team again realized that resolving the uncertainty would require user studies and market research, which the product manager estimated would cost at least $600K. A user study was listed as a constraint since the budget was frozen for the current fiscal year.

Competitive Advantage. Based on information gathered from attending trade shows and reviewing trade magazines, the team found no evidence of competitors integrating smart technology into this product category. However, competitors were using the technology in other applications and categories.

Value Proposition. The benefit of a small predictive and preventive device sounded compelling to the team, but they didn't know if it was a viable value proposition to the customer. The team determined that the value proposition could be explored further when conducting a user study to better understand the customer's perspective.

Profit. The team thought Smarty could be profitable, but further information needs to be collected through market research and user feedback to gauge commercial interests and general predictions: Will anybody buy? If so, how many? A factor supporting the profitability of Smarty would be the new recurring revenue stream for the failure prevention service.

Commercial Feasibility Pod Implications

The user studies were identified as a major constraint due to budget limitations during the current fiscal year. The user studies would have a ripple effect in determining market size, willingness to buy, unmet needs, value proposition, and product profitability. The team also realized that the user studies would be a key component of determining Smarty's technical feasibility. Still, the team was positive that Smarty could be very profitable, especially with the added recurring revenue that the 45-year-old product didn't have.

External Factors Pod

The External Factors Pod includes prompts on how the company adapts to factors outside the company. External factors are critical for companies that develop products in highly regulated and hypercompetitive industries. These factors are usually outside the control of most companies unless the company has a significant ability to influence government policy and regulations. Evaluating these factors can be challenging for a team since they are continuously in flux, as in the case of the economic climate, industry trends, and government regulations (Figure 5A.3).

Figure 5A.3: Cells in the External Factors Pod

External Factors Pod

How do we need to learn and adapt?

Industry Trends

Demographics

Government/ Regulatory/ Legal

External Factors

Technology

Economy

Social

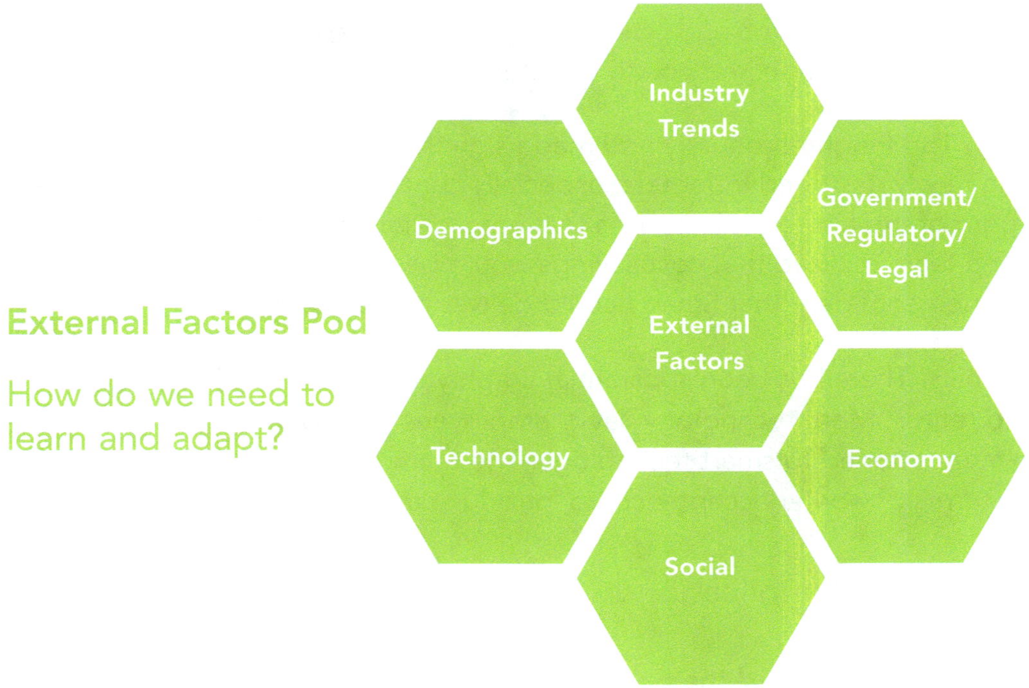

Industry Trends. The product manager kept abreast of industry trends by reviewing various freighter industry publications. A significant trend the manager identified was increased piracy. The team theorized that the uptick in maritime piracy might decrease the use of freighters. Another theory was that the owners of freighters might divert resources from control room upgrades to tighter security measures. Due to the uncertainty over the response to piracy, it would be essential to determine how these customers prioritize their spending. The team decided to add this issue to the user needs studies.

Government/Regulatory/Legal. Although the team thought regulations would not impede Smarty, the members decided to perform further due diligence and investigation. The following steps would include talking to industry contacts and the internal legal department.

Economy. The team determined that the economy would have a direct effect on whether the customer would buy Smarty. If there were a downturn in the economy, this would translate into fewer freighters transporting cargo. Freight owners would be even more vigilant in reducing costs and increasing efficiency.

The only way that Smarty could be successful would be if the technology were so important to customers that they would buy it regardless of a downturn in the economy. Smarty would need to demonstrate that it could significantly reduce operating costs through failure prevention, assuming owners of freighters made that their number-one priority. This assumption would be validated during the user study, and the team believed Smarty could deliver on this promised cost reduction.

Social. The team identified that various environmental groups were very concerned with freighter emissions and pollution and advocated greener transportation, like airships. The team did not identify this as an immediate threat.

Technology. The freighter industry had made significant gains in increasing efficiency and speed with the newly designed freighters, especially container ships. Smarty would align with the control room of the future, with devices that are smaller, connected, and enable remote monitoring. The team realized that cost savings would be essential for the owners of freighters, and they remained confident that Smarty could deliver on cost savings.

Demographics. There has been substantial employee turnover as the freighter workforce gets older and retires. More technology-savvy employees have been entering the workforce, and they are more attuned to smart technology. The team concluded that this would be a good time, in terms of worker acceptance, to introduce the product.

External Factors Pod Implications

The team once again identified the need for user studies. The studies would be imperative since the team needed to determine if Smarty was a priority over other technologies introduced to the operators. Team members were also confident that they could demonstrate cost savings once they got a workable prototype before the customer.

Although there has been talk of alternative modes of transportation, the team didn't see this as an immediate threat. The team concluded that the technology and worker acceptance of the smart technology was more of an opportunity than a constraint. Team members started to feel more upbeat and thought they could perhaps talk the management team into conducting the user studies within the year, especially since they were aware of how the budgets were padded.

Product/Technical Feasibility Pod

The next pod, Product/Technical Feasibility, investigates the company's ability to make a product that meets the intended customer's need, given the product and technology requirements and the company's resources. (Figure 5A.4).

Figure 5A.4: Cells in the Product/Technical Feasibility Pod

Product/Technical Feasibility Pod

Is the idea/product technically feasible?

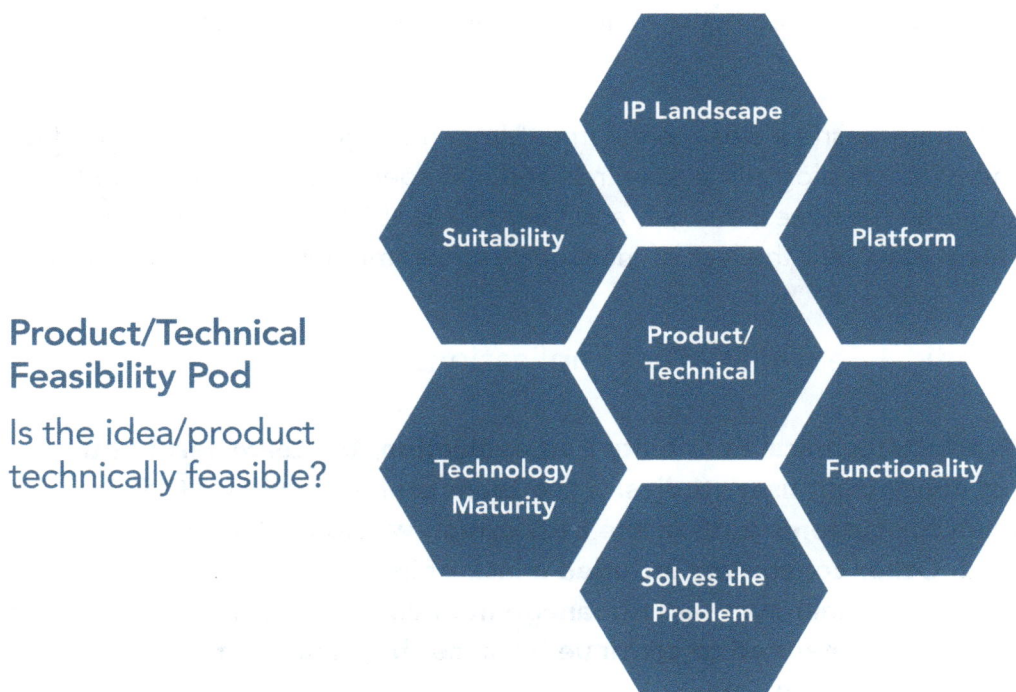

Intellectual Property (IP) Landscape. Before the meeting, team members reviewed existing patents to determine whether there would be any room for additional intellectual property (IP). The proposed technology was already used in other applications, so licensing the technology could be appropriate. The next steps included further investigation of the IP landscape after determining user needs. Also, the team needed to discuss the best option for the technology (license or patents) with input from the senior management team.

Platform. Team members thought they could treat this product as a beginning of a new platform. This would entail determining new extensions and whether the product could be modular to facilitate future development.

Functionality. Feasibility had been demonstrated in other applications, and the team was confident that the company could develop the product with the right resources. However, prototypes would need to be built and tested to reduce the functionality uncertainty.

Solves the Problem. Assessing the new product's ability to solve customers' problems would depend upon the results of the user needs studies.

Technology Maturity. The application of smart technology was new to this device. However, it was not a new technology used within the freighter. Consequently, the technology would be believable and credible to this customer base. The team was unsure if this technology would be disrupted in the near future, and they expected this technology to be the industry standard.

Suitability. There wasn't a clear commitment from management since the product required a drastic change in technology. The team realized management commitment to the project was needed, but the resources, budget, and internal capabilities to execute the project were limited. The team identified the lack of management commitment as another constraint.

Product/Technical Feasibility Pod Implications

During the Product/Technical Feasibility Pod evaluation, the team verified the importance of conducting user studies. If the team couldn't validate that Smarty would solve a problem for customers, then the best option would probably be to put the project on hold. Other constraints included a lack of internal resources to develop the smart technology and unknown senior-management commitment. The team members felt defeated, but they needed to continue because they had promised management a complete PRF evaluation on Smarty.

Business Configuration Pod

Recall that a company's Business Configuration refers to the resources, assets, activities, processes, technologies, capabilities, and choices that enable creating the new product. Too often, in our experience, companies take on a product development project, only to discover late in the process that the company is missing key business elements, such as the appropriate distribution channels, expertise in key technologies, or the ability to serve new customers.

The assumptions associated with the Business Configuration Pod include whether the business can support and leverage the proposed product. Exploring this pod helps the team consider the way the company creates and distributes a compelling value proposition that customers are willing to pay for at a price that yields an acceptable profit (Figure 5A.5).

Figure 5A.5: Cells in the Business Configuration Pod

Business Configuration Pod

Can the business support the idea?

Target Market

Key Partners

Channels

Business Configuration

Capabilities

Revenue Streams

Product Compatability

Target Market. The team realized that the new product would serve the company's existing target market (freighters), but the company needed to understand current unmet needs, problems, and drivers.

Channels. The team wasn't sure if the existing sales force could sell the smart technology because they had no experience in this technology. The business unit already had a direct sales channel to the owners of freighters, operators, and builders. Still, the move to the future control room might require a new sales channel to sell smart technology to retrofitters and builders of new ships. The salesperson on the PRF review committee indicated that for the company to overcome this hurdle, this channel would need smart-technology training.

Revenue Streams. The new product would generate a new revenue stream from charging an additional annual fee for the failure prevention service. Despite these added fees, the team realized that the company still must demonstrate cost savings to the customer.

Product Compatibility. The company's brand stands for quality and durability, but the products were stuck in the past, and the business unit was not known for being innovative. Smarty would be a departure for the business unit since the new product would combine innovation with quality and durability. However, this was compatible with the company's latest strategic intent on innovation, so the product aligned nicely. Also, the team thought the competitive advantage of the company's strong brand would encourage customers to try the new product.

Capabilities. Since the team had no experience with this smart technology, they would need to determine how to obtain the right intellectual property (outsourcing, licensing, building internal capabilities, or acquiring it). Smart technology would require new test equipment, manufacturing facilities, and resources with smart-technology expertise. It would also require extensive training of sales and service employees. If they developed the product internally, the team thought this would affect the product development process, since the business unit had not integrated software into its past products. They discussed a separate Skunk Works group as an optimal way to develop Smarty. Other considerations had to be given to selecting a department to oversee this type of smart technology in the future, not only for the business unit but also for the entire corporation.

Key Partners. The team identified the need for new suppliers, which would depend on where the technology would originate (internally or from a license). The team also indicated that the business unit might consider buying a company with smart-technology capabilities or forming a partnership.

Business Configuration Pod Implications

During the evaluation of the Business Configuration Pod, the team identified the major internal constraints for Smarty. Not surprisingly, the team again identified the user studies as a constraint for understanding the customer base due to the budget required. Other constraints included extending the sales and service channels and the cost incurred with new manufacturing capabilities, testing equipment, sales training, and internal resources. The team was also concerned that developing Smarty internally would be disruptive to the organization. Still, team members thought they could overcome this if they developed Smarty within a Skunk Works structure.

Wrap-up: The Product Risk Framework Management Review

Although the product manager and engineers were initially major supporters of Smarty, going through the PRF review tempered their enthusiasm. The perspectives of the other team members gave them a more realistic view of the project.

The PRF evaluation made the team more confident when discussing Smarty with the management team. They achieved this confidence by working together as an integrated team within a two-and-a-half-hour meeting. Team members discussed the opportunities, constraints, and uncertainties, giving participants a holistic view of the product. They also provided recommendations on which assumptions were most important to address and proposals for the next steps. One of the most significant constraints was the current budget, which made the proposed user studies impossible.

Fortunately, the meeting with management went well, mainly because of the team's open culture and objective discussion of the findings. Management agreed with the importance of the user studies and recommended that the team phase the studies and the cost. Getting this green light from the executives allowed the team to investigate the most critical uncertainty: owners' needs and priorities. Clearing this uncertainty helped the team determine the project's commercial viability and refine the value proposition.

The team used the Product Risk Framework Assumptions Tracker to track the project's uncertainty reduction over time. During month two, the team added additional assumptions, but they were able to substantially reduce the level of uncertainty over the next eight months (Figure 5A.6).

Figure 5A.6: Project Certainty Profile Over Time

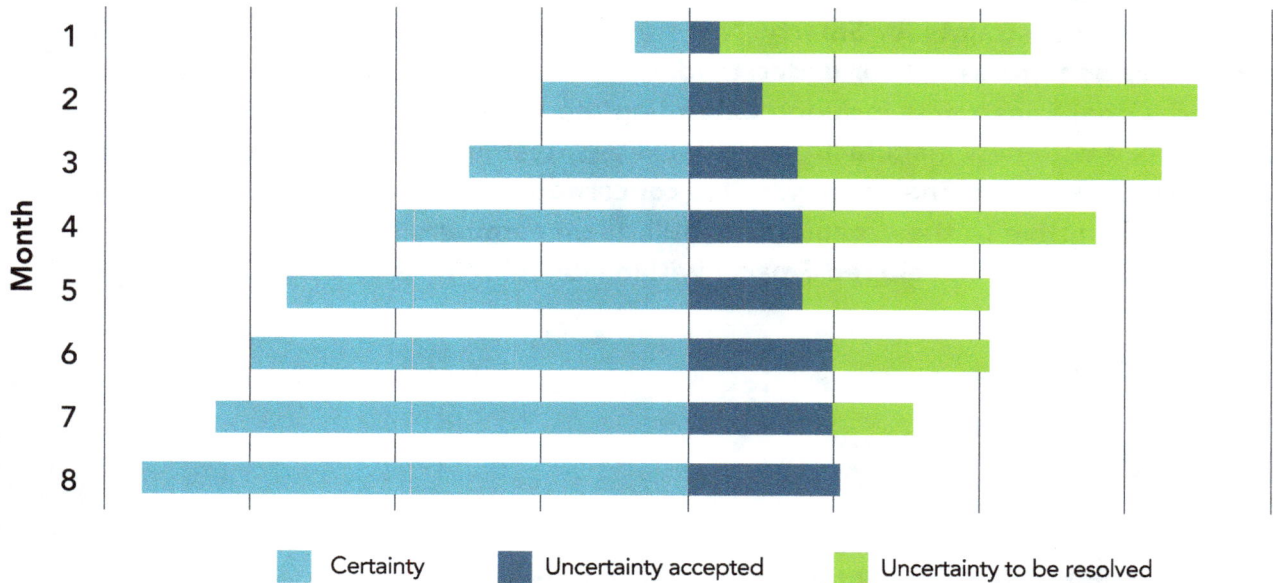

In closing, the case of Smarty illustrates how the Product Risk Framework can help companies manage the complex nature of product development in a changing environment. As the business unit managers realized, constraints in a changing environment are not always known and static. This creates uncertainty that must be explicitly considered in product development.

Notes

1.   C. Merle Crawford and C. Anthony DiBenedetto, *New Products Management* (New York: McGraw-Hill/Irwin, 2000).

Chapter 6
Plan

Chapter 6 Contents

What to Expect

Upon completion of Investigate (described in Chapter 5), the cross-functional team has prioritized the most critical assumptions, determined the maturity of the idea, and formed a recommendation on whether the new product idea merits a further investment.

This chapter explains the planning function in ExPD and how planning provides the flexibility, adaptability, and speed needed in product development. In ExPD, "plan" does not mean something fixed. Planning is flexible enough to adapt to many factors, including external and internal changes. Also, planning is not a one-time activity; instead, it applies throughout the entire process (Figure 6.1).

Figure 6.1: Planning in the Explore & Create Segment

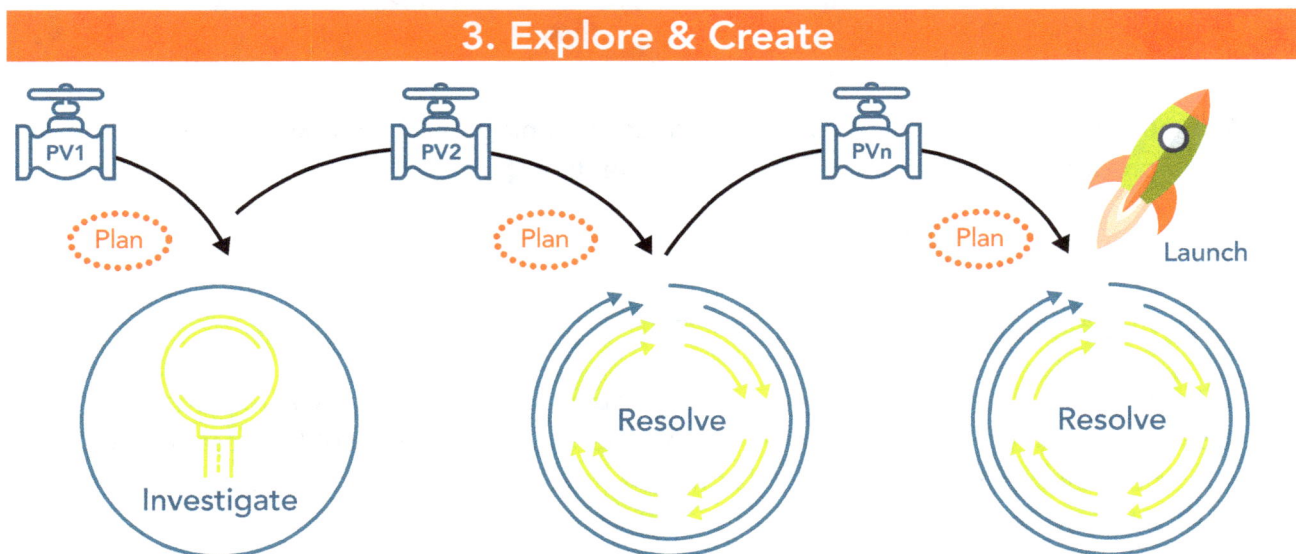

We then look at the tiers of activity involved in planning, including creating the Product Charter and Resolve Development Plan (RDP). These two documents provide the infrastructure to support resolution and development activities. We cover Prioritization Valve 2 (PV2), which includes deciding whether to continue with the product. We conclude the chapter by discussing the importance of reviewing the product portfolio.

Planning for Uncertainty: Flexibility, Adaptability, and Speed

New product development projects become more uncertain and riskier as companies navigate new markets, technologies, and products. Planning for high risk projects can be challenging. This chapter will discuss essential planning principles to help companies navigate uncharted territories, including the importance of flexibility, adaptability, and speed.

Planning for projects in uncharted territories is challenging.

With new markets, technologies, and products, planning must be **flexible** to accommodate each new product's unique nuances and requirements.

Planning must also be **adaptable** to handle uncertainty. The product development team adapts project activities and product definition as changes in the environment occur and new learnings are gathered throughout the project.

Lastly, **speed** enables the enterprise to reap the new product's financial rewards as soon as possible. In addition, speed allows the enterprise to enhance competitiveness by satisfying the customer's evolving needs. Speed also means reduced costs, including resources, development, and opportunity costs.

ExPD offers flexibility, adaptability, and speed by leveraging several proven principles:

- **Flexibility** is achieved through an infrastructure that provides direction, accountability, and tools for planning and executing the project

- **Adaptability** is achieved through iterative planning and a feedback mechanism for incorporating change and learnings into planning

- **Speed** is achieved by applying various lean, agile, and ExPD principles

Flexible Planning

With ExPD, the flexibility of the product development process is enabled by the two tracks described in Chapter 5, "Investigate": one for uncertainty resolution and the other for development (see Figure 6.2). Track A identifies the uncertainties that must be resolved to manage the project to an appropriate risk level. Track B identifies the Idea Maturity Model (IMM) activities that need to occur to advance the definition, design, and development of the new product.

Figure 6.2: Simultaneously Resolve & Develop

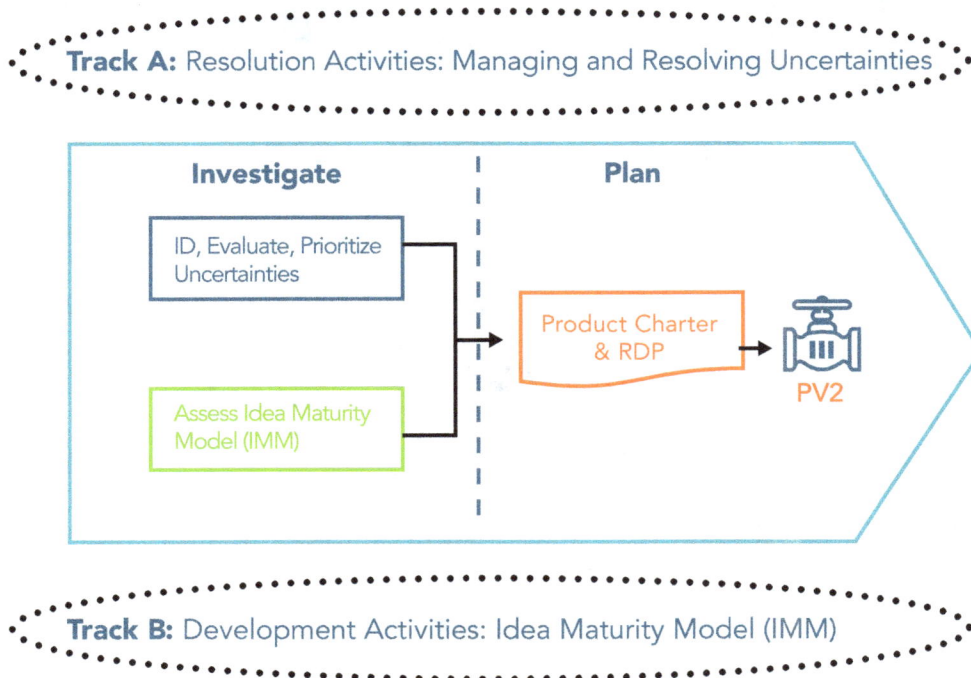

Track A: Resolution Activities: Managing and Resolving Uncertainties

Investigate | Plan

ID, Evaluate, Prioritize Uncertainties

Product Charter & RDP

PV2

Assess Idea Maturity Model (IMM)

Track B: Development Activities: Idea Maturity Model (IMM)

These two tracks are closely integrated and communicate the dual priorities of reducing uncertainty and developing the product. Throughout the Explore & Create segment, the product development team determines, plans, and executes the activities of Track A and Track B until the product is launched. Each step is an opportunity to modify or flex the process as needed.

The flexibility of the product development process is enabled by the two processes of reducing uncertainty and developing the product.

Adaptable Planning

In ExPD, iterative planning with quick iterations gives developers the flexibility to continuously adapt to changes and learnings. Therefore, we developed a tiered approach to planning, as shown in Figure 6.3. The tiers are the Product Charter, Resolve Development Plan (RDP), and Team Execution.

Figure 6.3: Adaptable Planning

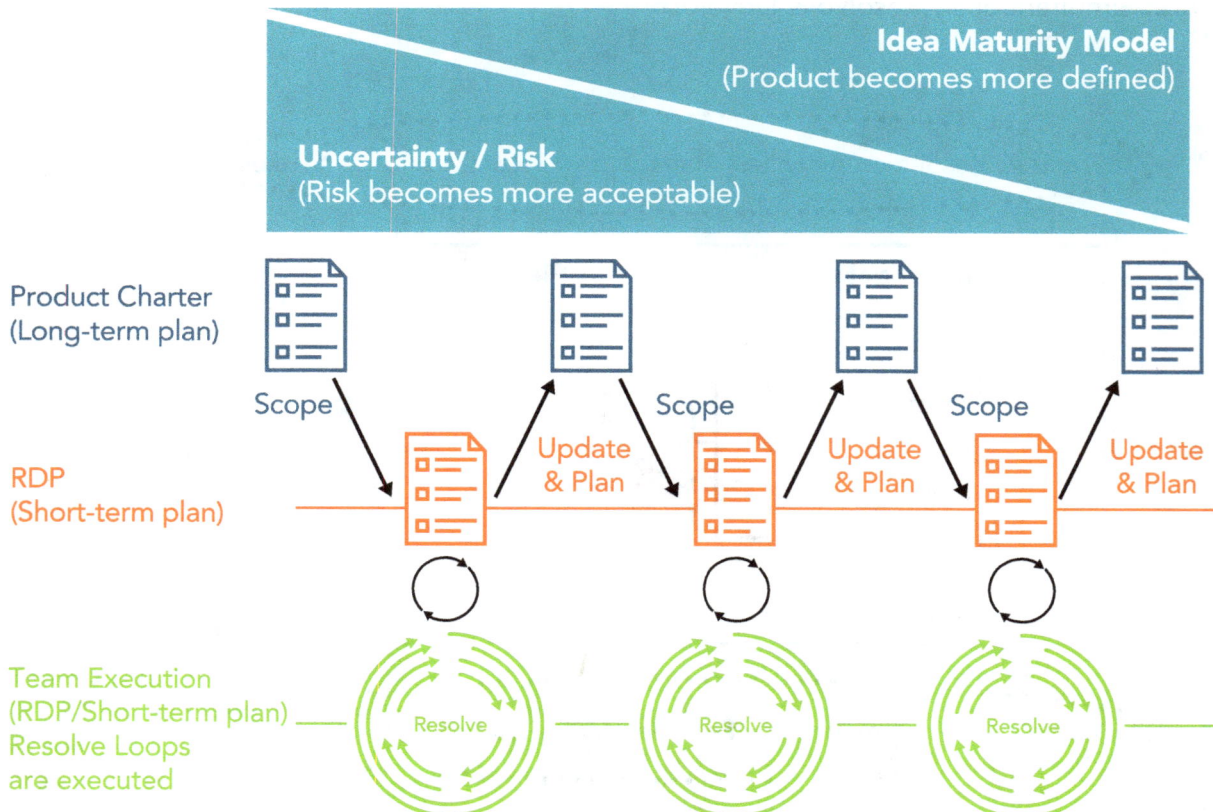

The product development team cycles through these tiers, as shown in Figure 6.4. The Product Charter and RDP are aligned. The Product Charter represents the long-term plan for the product. The RDP provides short-term planning and is designed to direct the team during execution. Team execution provides feedback for revising the product as the team gains knowledge.

Reinertsen highlights the important military practice of balancing centralized coordination with decentralized execution.[1] Our adaptable planning approach is based on this practice. The RDP helps support decentralized team execution, while the Product Charter provides coordination and a centralized vision for tackling the project.

Figure 6.4: Information Flow

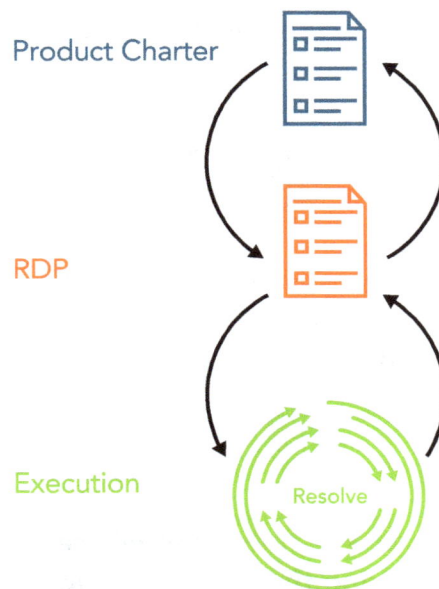

Product Charter

RDP

Execution

Resolve

Product Charter

The Product Charter provides the overarching long-term plan for the product; it includes the vision and goals for the final product. The charter's scope is driven by the prioritized assumptions, and development activities outlined in the Idea Maturity Model (IMM). Initially, the Product Charter informs the RDP and team execution.

Resolve Development Plan

The Resolve Development Plan (RDP) provides short-term planning—two to four weeks—for iterations at the team execution tier. A quick recap:

- Often companies use the terms iteration and sprint interchangeably. For simplicity's sake, we will use the term iteration throughout this user guide

- Iteration is a specified unit of work that must be completed to reach the desired goal quickly

- We use the term Resolve Loop for the resolution of uncertainty that can be achieved through iteration(s)

- Multiple iterations can run concurrently if the assumptions are independent

Resolution of uncertainty can be achieved through iterative learning.

The RDP provides the project team with guidelines to resolve uncertainty during project execution and goals for development aligned with the IMM. It is a focused and straightforward document so the team can adapt quickly to product uncertainties.

Team Execution Tier

The Resolve Loops occur within the Team Execution tier where uncertainty and risk are resolved. The team gathers learnings during the iterations and feeds them back into the RDP and the Product Charter.

The tiers may sound time-consuming, but in fact, the flexibility achieved is well worth the effort. Over the years, we have encountered some common planning misconceptions. Sidebar 6.1 summarizes them and explains why a good approach to planning pays off.

Sidebar 6.1. Fail to Plan, Plan to Fail

The adage "Fail to plan, plan to fail" holds true with ExPD. Planning is critical to effective project execution. The lack of investment in planning leads to insufficient learning and, ultimately, product failure.

We have worked with companies that see project planning as a necessary evil. Common excuses for the lack of planning include the following:

- "Too much uncertainty makes predicting impossible, so why waste your time?"

- "Time spent planning is taking away from doing."

- "We don't have the time to plan; we have too many other irons in the fire."

- "It's going to change anyway."

Rigid, frozen plans are toxic to innovation and learning. ExPD adopts a novel approach to planning that adapts project activities as new information is acquired and the product evolves. You may have heard the expression, "No battle plan survives first contact with the enemy."[2] The point is that a plan provides a starting point that changes and evolves based on what happens on the battlefield in real-time. The value is not in the written plan itself, but in the analysis, preparation, and adjustments as information is learned.

In a dynamic world, markets and technology are forever changing. These changes and newly acquired knowledge become the catalyst for adjustments to the plan. This allows the project to adapt to external and internal forces.

Speed in Planning

ExPD achieves speed by integrating these practices and principles:

1. Adapt quickly to learnings and changes in the environment and streamline decision-making through strategy and roadmaps.

2. Eliminate unnecessary activities. The flexible nature of ExPD means that the project team determines what activities are necessary based on the product's unique nuances and requirements.

3. Eliminate unnecessary bureaucracy, with management by exception. The team is self-governing to deliver on the project. If the team is on track to deliver, team members are not interrupted by unnecessary meetings.

4. Manage the project in successive iterations, reducing batches of work and decreasing cycle time.

5. Identify and resolve the most critical uncertainties early in the process, so projects fail fast and fail early, or adapt.

6. Minimize documentation. The nature of the uncertainties and development activities dictates what documentation is necessary. At minimum we recommend:

   - Product Charter and RDP

   - Product Viability Analysis (Sidebar 5.6) before hitting big spend

   - Risk management requirements and systems for regulated industries

   - Idea Maturity Model (IMM) to support the ongoing development of the product

   - Modular documentation focused on the unique nuances of the product instead of a generic gate document

7. Constrain the number of projects with the Prioritization Valves (PV) to match available resources. This results in improved throughput time.

Next, we will do a deeper dive into the components of the Product Charter and Resolve Development Plan (RDP).

Product Charter

The Product Charter and Resolve Development Plan (RDP) provide the infrastructure to support resolution and development activities, depicted as Track A and Track B in Figure 6.5. The Product Charter includes a high-level plan to resolve the prioritized uncertainties and informs the more detailed RDP activities.

Figure 6.5: Product Charter & Resolve Development Plan

Track A: Resolution Activities: Managing and Resolving Uncertainties

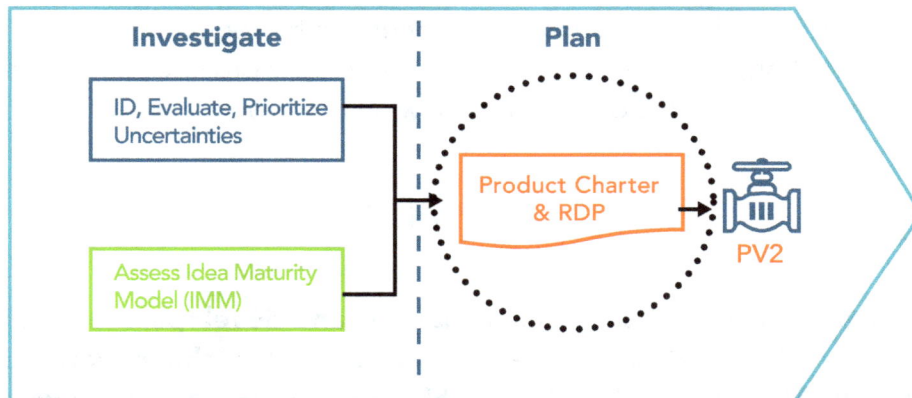

Track B: Development Activities: Idea Maturity Model (IMM)

Developing the Product Charter

Our Waze example from Figure 1.3 in Chapter 1, "The Case for ExPD," also applies to how an ExPD project is managed. A way to think of iterative planning is to imagine getting ready to take a long road trip. You might use Google Maps® or Waze® to find the major roads to travel to your destination. Then as you get nearer to specific intersections or run into unexpected road construction, you zoom in and plan the details. Similarly, the Product Charter is initially the high-level roadmap of how your project will proceed, and the RDP provides the detail as you approach your destination.

The Product Charter is initially the high-level roadmap of how your project will proceed, and the RDP provides the detail as you approach your destination.

The Product Charter is a living document that needs to be regularly updated to reflect the project's latest information, activities, and resources. It initially provides a target that is not specified in features and attributes but in terms of what the product needs to accomplish.

Components of the Product Charter

The Product Charter thoroughly captures managers' and developers' vision of a successful product without a detailed specification subject to change later as the exploration unfolds. The Product Charter is a high-level document that evolves as uncertainty is resolved and the product idea matures. It includes the following six key components (Figure 6.6):

Figure 6.6: Major Components of the Product Charter

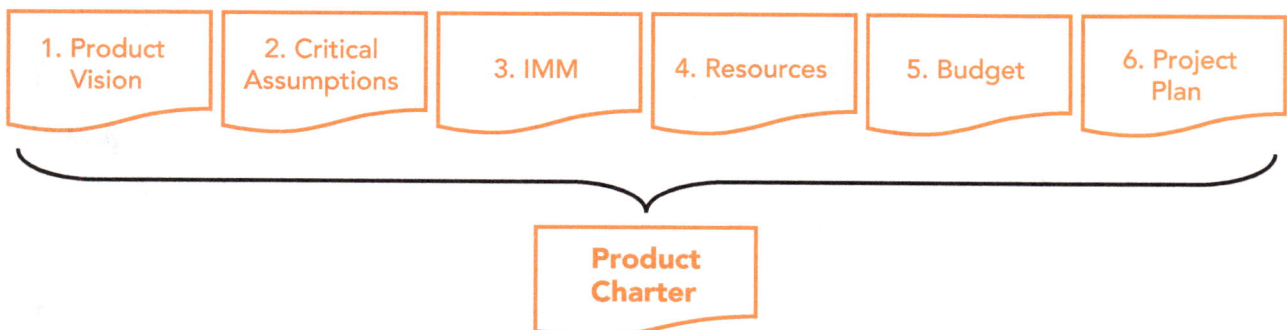

1. The **product vision** is a high-level overview of what the product will do, who it will serve, and how it will differ from what your competitors are currently doing.[3] A vision is like a compass used on a trail; although the trail wanders, you check the compass to ensure that you are going in the right direction.

2. During Investigate, the team identifies, evaluates, and prioritizes the most **critical assumptions.** There should be an understanding of the types of resolution methods and the level of effort needed to resolve these assumptions at a high level.

3. High-level product development activities and goals are based on the **Idea Maturity Model (IMM).** Using the IMM, the project team identifies the high-level development activities to move the product toward launch.

4. High-level **resource** requirements are captured within the Product Charter. This list may include key tasks, resources to execute the tasks, and handoffs or interactions between the disciplines.

5. The **budget** includes funds for all resources and the projected cost for the entire project. Cost is truly projected, and the budget can change based on learnings during project execution. Therefore, the project budget must be continuously reviewed and updated throughout the project. Capturing assumptions that drive the budget makes this update process easier and transparent. Here, it may be helpful to introduce a reverse income statement (see Sidebar 6.2). This is a valuable tool for tracking financial viability.

6. The projected **project plan** and time frame like the budget are projections for the entire project based on major milestones. This timeline reflects current information and gets updated as the team executes the project.

Sidebar 6.2. Reverse Income Statement

Ultimately, you want to determine whether the expected profit will make the product worthwhile before proceeding with the project. Unlike the traditional approach of estimating a product's revenues and then assuming profits will follow, McGrath and MacMillan propose creating a Reverse Income Statement when developing a new product/venture where uncertainty lurks.[4] The authors outline five critical steps:

1. Bake profitability into your venture's plan.

2. Calculate allowable costs.

3. Identify your financial assumptions.

4. Determine if the venture still makes sense.

5. Test assumptions at milestones.

Any critical variables that may have an impact on financial results should be folded into the financial plan. This may give rise to a new uncertainty. For example, discovering that manufacturing cost has increased may trigger an uncertainty about whether customers are willing to pay more for your product.

To read Turba's Product Charter for its PERS product, see Sidebar 6.3.

Sidebar 6.3. The Turba Team Develops the Product Charter

TURBA CORPORATION

The Turba cross-functional team, including the project and management teams, met to review the Product Charter. The project manager was instrumental in pulling together the different components of the charter.

Product Vision

Turba's product manager developed the product vision based on the strategies outlined in Chapter 3, "Why Strategy Matters for Product Development." Then, the proposed vision was presented to the cross-functional team, and they worked together in fine-tuning it:

"This is a premium PERS product for seniors living alone that works anywhere. We will use available technologies that have already been demonstrated to be commercially and technically viable. We differ from our competitors in that we provide the highest-quality products and services."

The cross-functional team agreed that the vision provided the necessary high-level boundaries for developing the product.

Critical Assumptions

During Investigate (Chapter 5), the following two assumptions were prioritized as the most critical to resolve.

1. Target users living at home that are willing to pay for a PERS that works everywhere, including outside the home.

2. Bluetooth technology will reliably enable PERS to work anywhere your smartphone works.

Idea Maturity Model

In Chapter 5, "Investigate," the Turba team identified PERS as an Idea on the IMM, and the objective is to reach the Stated Concept level within three weeks.

A Stated Concept includes the starting point for research and experimentation to discover possible solutions (see Table 6.1).

Table 6.1: ExPD Idea Maturity Model

	Maturity Level	Description	Key Activities & Tools	Launch Activities
START	**Idea**	Initial thought on how to address the opportunity	Brainstorming, customer solution, problem solving, competitive analysis, etc.	None
FINISH	**Stated Concept**	Potential solution(s) to address the opportunity/ problem to be solved, plus a clear statement of benefits. Hypothesized value proposition. This is the gateway to research and experimentation	Prototypes, artist renderings, Minimum Viable Product (MVP)	None

Resources

A full-time equivalent (FTE) product manager was assigned to help drive down uncertainty on both critical assumptions, but the primary emphasis is on Assumption 1 (Willingness to Pay).

Three FTEs, including an experienced engineer, a junior engineer, and an outside Bluetooth/electrical engineering consultant were assigned to Assumption 2 (Bluetooth Technology).

A half-time equivalent project manager was assigned to oversee the progress and learnings on the two critical assumptions.

Budget

The budget for resolving the two critical assumptions is $150,000.00, including salary and other resources, such as lab time and equipment. This budget may change as the project team executes iterations and learns more.

Project Plan

The project manager drafted a high-level plan outlining the major milestones based on the two most critical assumptions and IMM activities.

Resolve Development Plan (RDP)

We coined the term Resolve Development Plan (RDP) since it is essential for the team to resolve uncertainties and develop the product. The exploratory nature of ExPD requires latitude for the team to adapt to discovery and learning. At the same time, uncertainties are being explored in the Resolve Loops (which will be described in Chapter 7, "Resolve"). To maintain this flexibility, the RDP relies on short iterations.

In preparing for PV2, the team drafts the RDP to define what needs to be done to reduce uncertainty for a given assumption and move the product along the IMM. As the team runs the Resolve Loops, the RDP is updated based on the new information and learnings gathered during each iteration. The updates should include any short-term changes in the project, including uncertainties that were resolved, not resolved, or newly discovered.

Each iteration's duration is based on what the team needs to learn or resolve, though typically, it is in the range of two to four weeks. On the one hand, iterations should be short enough to ensure that important discoveries are surfaced quickly, so time is not wasted pursuing blind alleys. On the other hand, iterations should be long enough to produce meaningful results. A good rule of thumb is that iteration duration should be about as short as it might take to resolve one simple assumption. Conversely, complex assumptions are divided into multiple iterations.

We recommend breaking the activities into small increments, which helps the team to be focused, fast, and budget-conscious. For example, global consumer product testing is expensive, but a three-week iteration might test locally with a small ethnography study. This first iteration does not eliminate all uncertainty, but it does provide confidence that the test methods are valid and would surface any major issues with the product concept.

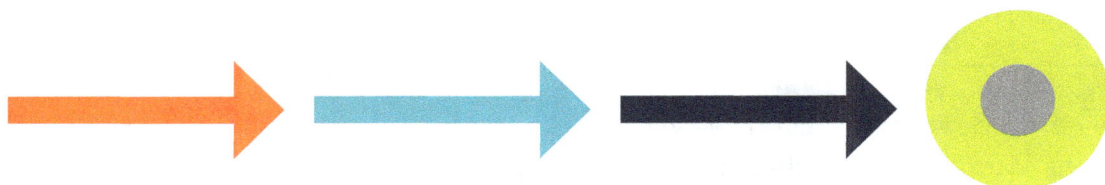

Companies that need to develop hardware and software often find that software-related iterations can be much faster than hardware-related iterations. Some companies run software at one speed and hardware on a longer time frame. They then build integration time and testing at 6- or 12-week intervals. Ultimately, it is up to your team to decide each iteration's timing.

Components of the RDP

Your RDP document should capture most of the components listed in Table 6.2. Feel free to add any other components that may be helpful for your organization.

Table 6.2. Components of the Resolve Development Plan (RDP)

Assumption description
Hypothesis to test
Learning goals
Nature of study/experiment
Metrics to track the resolution of uncertainty
Iteration disposition and date—indicates the final disposition of each iteration (Continue, Adapt, Cancel, or Hold) and an explanation for each decision and outcome
Assumptions Tracker—indicates uncertainty to be resolved during each iteration, detailing the approach(es) to resolve with specifics, updating progress or setbacks, and recording newly discovered uncertainties and risks
Pledge parameters, including budget, duration, scope, objective, and resources
Vision—indicates whether the product idea is still on vision/track
Current and target placement in Idea Maturity Model (IMM)—indicates the project's current placement on the IMM and how the goal will be reached
Detailed short-term timeline/project plan for iterations or project plan milestones and any dependencies
Next steps/project decisions (Explanation)

In the remainder of this section, we will drill down into three areas of the RDP: Pledge, Assumptions Tracker, and detailed short-term timeline or project plan.

Pledge

The Pledge is a bilateral agreement between the project team and senior managers that specifies the project's budget, resources, and time. It gives the project team a solid framework of goals and boundaries. Much like the Simple Rules described in Sidebar 3.2 of Chapter 3, "Why Strategy Matters for Product Development," the Pledge gives the team simple, straightforward, and tactical rules it can easily follow to respond quickly to fast-moving opportunities.[5]

ExPD developers do not report progress against frozen timelines, milestones, and budgets; instead, they keep a vigilant eye on the Pledge as they explore and resolve uncertainties. The Pledge acts as a contract, and both parties are expected to abide by its stated requirements.

The Pledge acts as a contract, and both parties are expected to abide by its stated requirements.

Although you can add additional parameters to the Pledge, we recommend at a minimum that you use the triple constraint parameters (resources, scope, and time) depicted in Figure 6.7. Within those constraints, the team agrees to an IMM objective and to reduce uncertainty for a given assumption(s) with specified resources, budget, and time.

Figure 6.7: Pledge Parameters

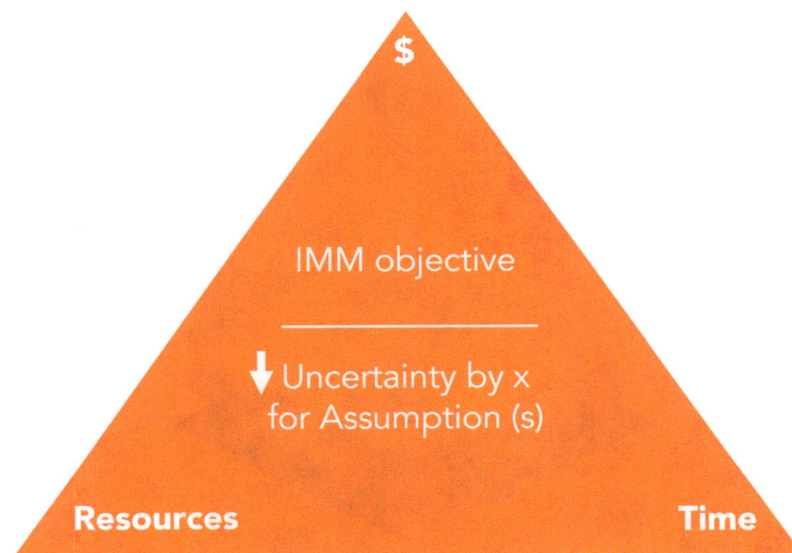

$
\text{IMM objective}
$
_____
↓ Uncertainty by x
for Assumption (s)

Resources · Time · $

If the team expects to exceed the stated parameters, it is responsible for flagging these issues. For example, suppose the team sees it will go beyond the budget or timeline for resolving an assumption. In that case, the situation is identified as an exception and handled using the principles of management by exception (explained in Chapter 8, "People").

Managers and the team should gather for a thorough project review when any Pledge terms come into question. At such a review, attendees may decide to learn more and continue, cancel the project, or adapt to address a different opportunity that has been uncovered. Another option is to put the project on hold. This can happen if a new higher-priority project needs your resources or the technology needs further development. However, the team must be careful not to opt for a hold decision to avoid canceling the project.

Assumptions Tracker

Tracking assumptions is a dynamic process that needs continuous updating. Assumptions will be resolved over time, new assumptions will be added, and some assumptions may spiral into other areas of uncertainty. Keeping tabs on what has or has not been resolved and tracking performance overall can be challenging.

Therefore, we extended the Product Risk Framework tool's functionality beyond identifying, evaluating, and prioritizing the most impactful uncertainties. It also includes an Assumptions Tracker that the project team can use to track the resolution of uncertainties and risks from product ideas through launch, as initially discussed in Chapter 5, "Investigate."

Assumptions will be resolved over time, new assumptions will be added, and some assumptions may spiral into other areas of uncertainty.

Additionally, the tool includes a visualization component that displays informative graphics and dashboards (Figure 6.8). This graphic displays the project certainty, the uncertainty to be resolved, and the uncertainty accepted (since a goal of resolving all uncertainty would be unrealistic). The Product Risk Framework's ultimate goal is to help the management committee and project team make better product decisions.

Figure 6.8. Product Risk Framework: Project Certainty Profile over Time

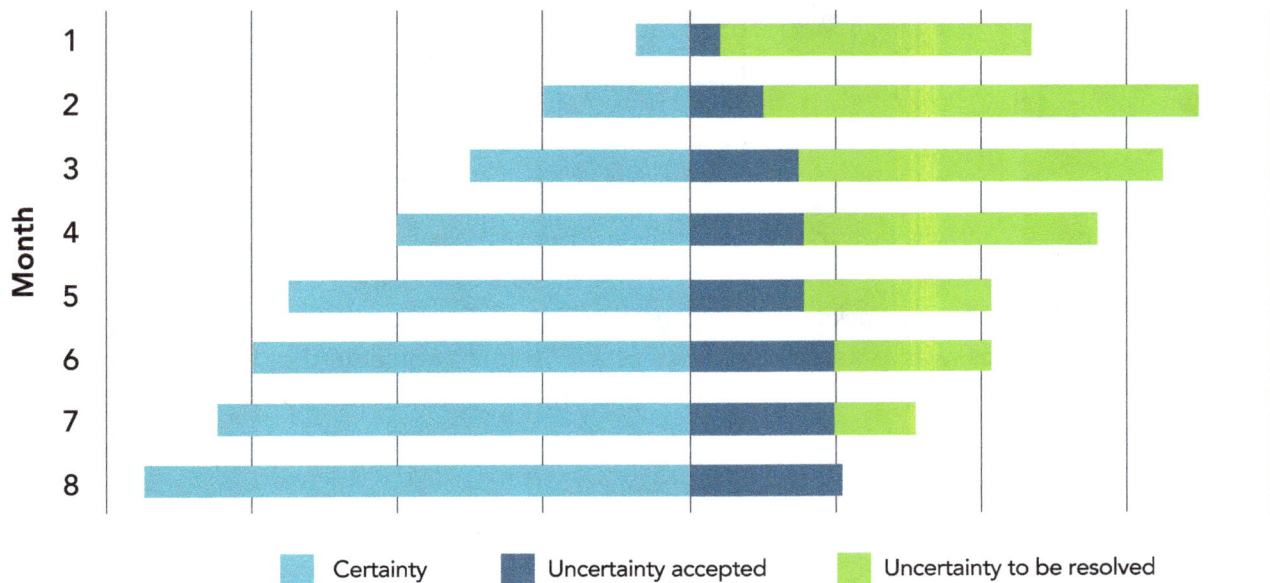

The team uses the Assumptions Tracker during iterations to record learnings, measure progress, and capture new or modified uncertainties and risks. It also stores all information related to each assumption, such as ratings, priorities, possible methods of resolution, learnings from each iteration, links to supporting documents, and relationships between assumptions.

If you don't want to use the Product Risk Framework tool with the embedded Assumptions Tracker, we recommend tracking assumptions within a spreadsheet. You can typically rank or bucket the projects based on impact (size of the potential loss, cost of resolving risk, probability of occurrence, etc.). Then, within the spreadsheet, collect your assumptions and track them. To get started, you can set up spreadsheet columns using the sample in Table 6.3

Table 6.3. Assumptions Spreadsheet: Sample Column Headings

Assumption Number	Category	Description	Detail Supporting Information	Risk Measure(s)	Priority	Status

If the team uses a spreadsheet, they may also want to use a risk burndown chart to record the overall number of risks being resolved on the project. A risk burndown chart is a graphical representation of the number of risks resolved (see Figure 6.9). The number of unresolved risks is on the vertical axis, with time along the horizontal. This is a valuable tool to keep your management team abreast of how the team is progressing to reduce the product uncertainty.

Figure 6.9: Risk Burndown Chart

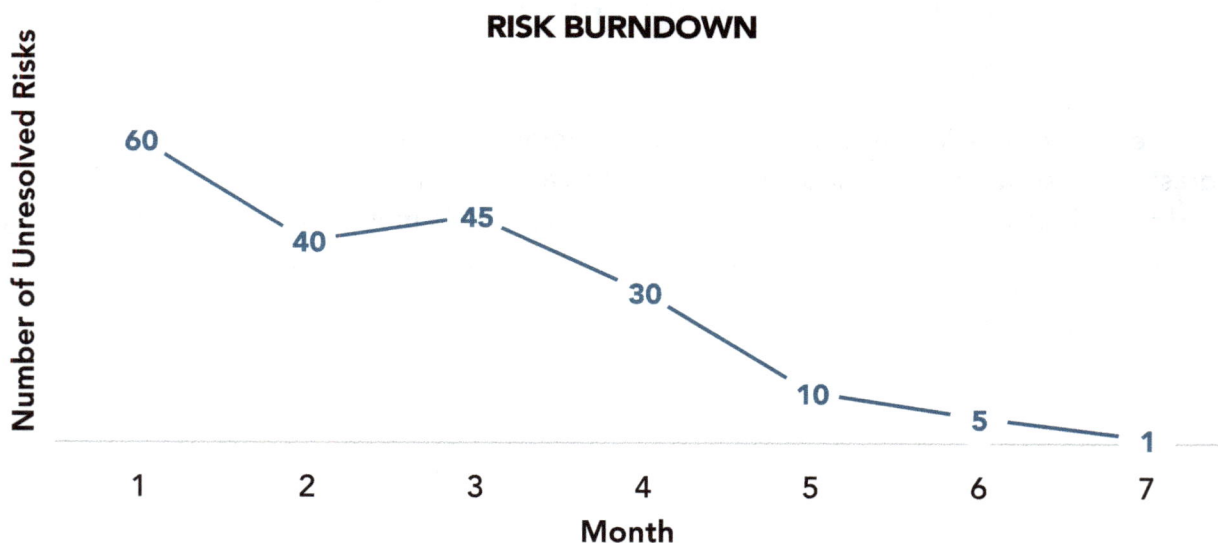

RISK BURNDOWN

Detailed Timeline/Project Plan

The RDP is a living and breathing document that includes a project timeline that the project manager and team must update while executing their resolution and development activities. As the RDP is updated with learnings, the project manager is responsible for updating any significant changes in the Product Charter. Therefore, the Product Charter and RDP should be in sync.

ExPD leverages principles from rolling-wave planning. The work to be accomplished in the near term is planned in detail within the RDP, while work in the future is planned at a higher level within the Product Charter. This section lays out how to use these principles in ExPD.

ExPD leverages principles from rolling-wave planning. The work to be accomplished in the near term is planned in detail within the RDP, while work in the future is planned at a higher level within the Product Charter.

Once the preliminary RDP has been created, refine the plan by adding details for two to four weeks, then break the planned work down to granular activities. This detail is used in agile and lean product development to build visibility and accountability. These activities should be estimated at one- to three-day durations, using named resources. Each week, the team should meet and complete the following tasks:

1. Update the current project plan by moving the start dates of any tasks that should have started to the anticipated start date and adjusting the duration of any tasks.

2. Add new tasks as needed, including resources, duration, and work (effort). Then, link these tasks in the project plan.

3. Upon completing an iteration, update the RDP with any changes in the project, including progress in the Idea Maturity Model, product vision, timeline and budget, and uncertainties that have been resolved, not resolved, or newly discovered.

The planning process can be complex for highly uncertain projects, but it remains essential. Sidebar 6.4 suggests ways to tackle planning for a project with high uncertainty.

Sidebar 6.4. Tips for Building a Project Plan for High Uncertainty

For highly uncertain products, you may not be able to specify detailed activities and durations, but you can use averages, ranges, and estimates in your plan. These will give you some idea of the level of effort and time frame necessary to complete the project.

We recommend the following tips:

- Include prioritized activities for uncertainty resolution. Then link them appropriately in the project plan so you can see any interdependencies

- Consider using the Reverse Income Statement approach to estimate potential revenue on a highly risky project (see Sidebar 6.2). Determine how much revenue you will need to deliver the required level of profits and the associated cost

- Understand the types of personnel and skill sets you will need

- When evaluating resource requirements, determine any potential overloads. This can be done by resource leveling, linking tasks sequentially, or determining a specific resource gap. Also, be sure to include physical resources, such as test labs, production equipment, and outside suppliers

- Keep the project visible by using visual project boards during 15-minute stand-up meetings

- Throughout the project, continuously update the RDP with the learnings discovered during execution. Update the Product Charter as appropriate

Sidebar 6.5. The Turba Team's RDP

TURBA CORPORATION

The cross-functional Turba team met to develop the Resolve Development Plan (RDP) for resolving the two most critical assumptions. Turba's RDP is illustrated in the next chapter (see Sidebars 7.2, 7.4, 7.6, and 7.7 on how Turba used the RDP during the Resolve process).

Now we go to the next step in the ExPD Process, PV2.

Prioritization Valve 2 (PV2)

Rather than building a highly detailed plan before PV2, you need to provide enough information to understand what needs to be accomplished to deliver the product, including the Product Charter and RDP. These two documents should give the management committee enough information during PV2 to determine whether the project should proceed (see Figure 6.10).

Figure 6.10: Inputs into PV2

Track A: Resolution Activities: Managing and Resolving Uncertainties

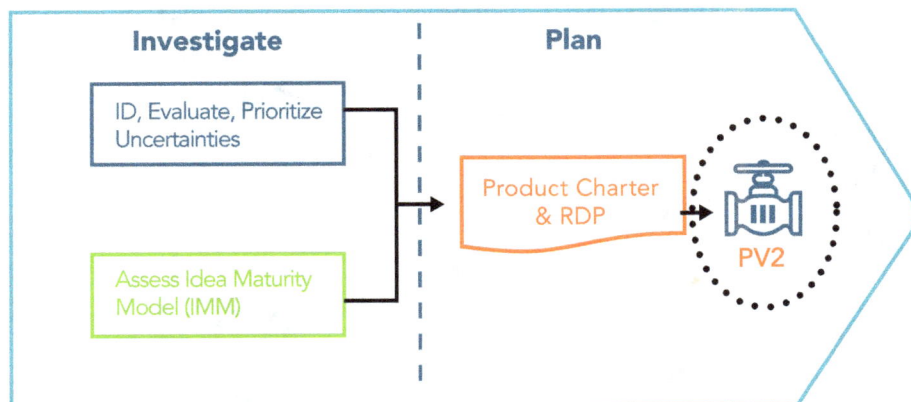

Track B: Development Activities: Idea Maturity Model (IMM)

If the project is chosen to continue, the management committee and team agree on the Pledge. They evaluate the availability of resources. If resources are not available, the management committee places the project in the queue until resources are available. When appropriate resources are available, the project is released to the next step within the process, Chapter 7 "Resolve."

Process Characteristics of Prioritization Valves (PV's)

Individual projects evaluated, scored, prioritized, and resources allocated

PV2 and the Prioritization Valves (labeled PVn in Figure 6.11) have similar process characteristics. At each PV, the management committee evaluates and prioritizes projects. The committee also manages and commits resources to prioritized projects.

Figure: 6.11. Prioritization Valves for Explore & Create

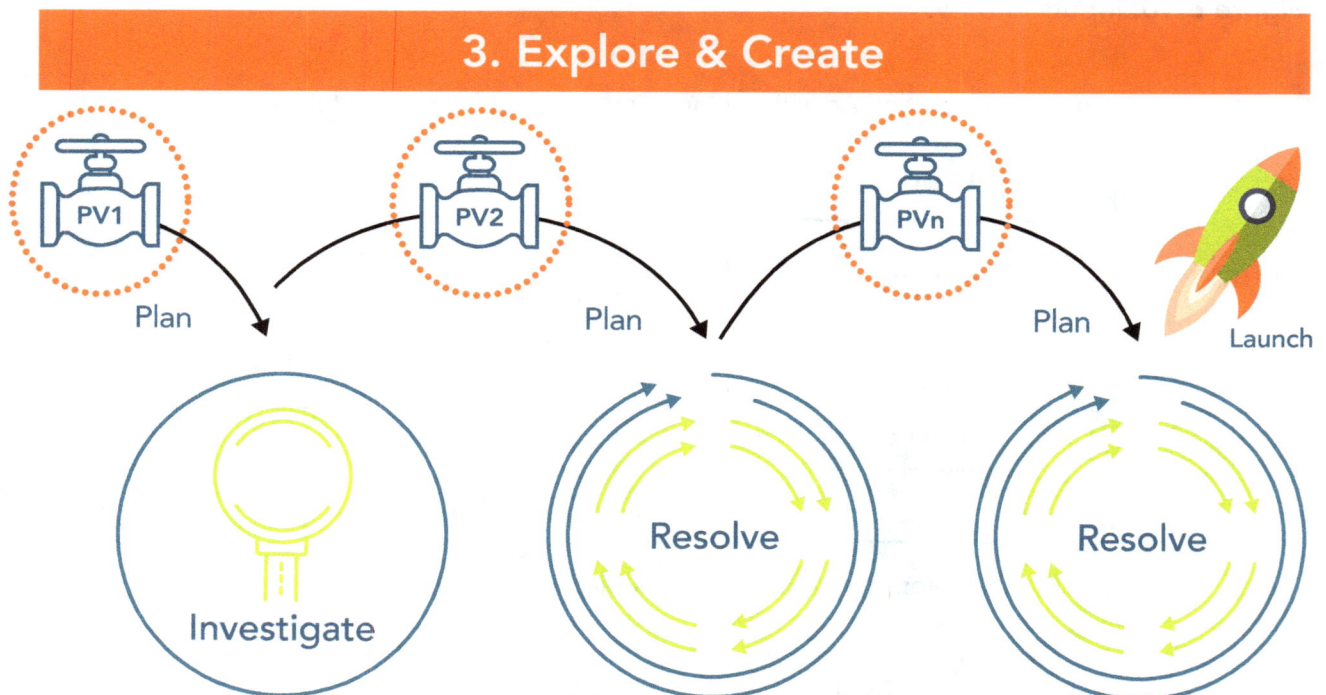

PV1, PV2 and the subsequent PVs have the same process characteristics, but there are some differences. PV1 utilizes the Idea Screening Package and acts as a high-level Prioritization Valve to determine which product ideas will progress to the Investigate process. PV2 and the subsequent PVs utilize the Product Charter and RDP as documentation for decision-making on an individual product investment.

During PV2, it is a good time for the management team to review the product development portfolio, as discussed in the next section.

Portfolio Management

A product development portfolio session provides management with a holistic view of the products currently in the portfolio, as opposed to a product development process focusing on one product at a time.

In a typical organization, portfolio management sessions evaluate the status of active projects to decide whether they should proceed, pivot, be discontinued, or be put on hold. The management committee also makes decisions about product ideas awaiting evaluation to enter the product development process. These decisions are generally based on the required investment, the expected benefit or return, strategic fit, and availability of appropriate resources. Ultimately, the management committee makes strategic and investment decisions impacting resource allocation.

The purpose of a product portfolio is three-fold: 1. maximize the value of the portfolio, 2. deliver on company strategy, and 3. maintain an appropriate balance on important dimensions. For example, a company may strive to allocate resources equally across product lines, or allocate a greater percentage of resources to particular market segments. A critical, but often underappreciated and overlooked dimension, is the risk of individual projects and the cumulative risk of the portfolio. Management needs to maximize the value of the portfolio while understanding the overall level of risk.

One way to approximate risk is by product type. Breakthrough product types typically represent the most risk since uncertainty is the greatest. At the other end of the spectrum, incremental product enhancements represent the least risk because uncertainty is lowest. As mentioned earlier, R&D spending has been shifting to newer and breakthrough products[6] since higher-risk products generally translate into higher returns.

To avoid such pitfalls, we integrated a portfolio view into the Product Risk Framework (PRF). It is a visual starting point for dialogue about the enterprise's mix of projects and their fit with strategy and risk tolerance (see Figure 6.11). It measures risks across the spectrum of technical feasibility, external factors, business configuration, and commercial feasibility.

The Product Risk Framework (PRF) portfolio view is a visual starting point for dialogue about the enterprise's mix of projects and their fit with strategy and risk tolerance.

Figure 6.12: Product Portfolio

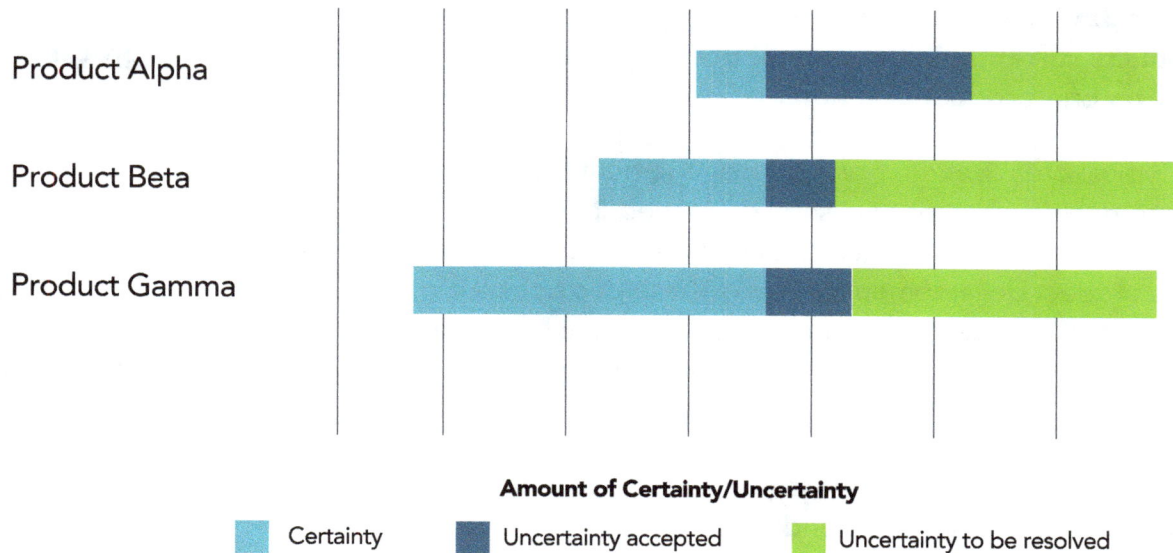

Amount of Certainty/Uncertainty

Certainty   Uncertainty accepted   Uncertainty to be resolved

A good starting point for this dialogue is at PV2. Upon reviewing the PRF portfolio view, the team may decide that the level of uncertainty is too high for some product ideas, such as Product Beta in Figure 6.12. Or perhaps a product idea has such high strategic importance that the company will proceed cautiously.

During each subsequent PVn, the team should continue to review the portfolio of projects. Then, within the PRF tool, the team can gauge the reduction of uncertainty across the projects within the portfolio.

If you want to learn more about portfolio management we recommend these resources.[7,8]

In Chapter 7, "Resolve," we will introduce you to the Resolve Loop.

Key Chapter Points

1. **Planning within ExPD provides the flexibility, adaptability, and speed needed in product development.**

2. **The Product Charter provides the overarching long-term plan for the product; it includes the vision and goals for the final product without a detailed specification subject to change later as exploration unfolds.**

3. **The Resolve Development Plan (RDP) provides short-term planning for iterations at the team execution tier. To maintain flexibility, the RDP relies on short iterations.**

4. **Uncertainty and risk are resolved by the team during iterations of the Resolve Loop. During iterations, the team gathers learnings and feeds them back into the RDP and Product Charter.**

5. **The Pledge gives the project team a solid framework of goals and boundaries. A bilateral agreement between the project team and senior managers specifies a budget, resources, and time.**

6. **The Assumptions Tracker maintains the current status of all assumptions and provides a yardstick to track risk reduction. You can potentially identify many assumptions within your project, so having a way to manage and track them is very important.**

7. **ExPD leverages principles from rolling-wave planning. The work to be accomplished in the near term is planned in detail within the Resolve Development Plan (RDP), while work further in the future is planned at a higher level within the Product Charter.**

8. **PV2 is an excellent time to review the portfolio of product ideas. Upon reviewing the portfolio, the team may decide that the level of uncertainty is too high for some of the product ideas.**

Notes

1. Donald Reinertsen, *The Principles of Product Development Flow: Second Generation Lean Product Development* (Redondo Beach, CA: Celeritas Publishing, 2009).

2. Helmuth von Moltke the Elder, Prussian field marshal (1800–1891). Several English translations have been published, with slight differences in the wording.

3. Preston G. Smith, *Flexible Product Development: Building Agility for Changing Markets* (San Francisco: John Wiley and Sons, 2007; reprint 2018).

4. Rita Gunther McGrath and Ian C. MacMillan, "Discovery-Driven Planning," *Harvard Business Review* (July–August 1995); Rita Gunther McGrath and Ian C. MacMillan, *Discovery-Driven Growth* (Boston: Harvard Business Press, 2009).

5. Kathleen M. Eisenhardt and Donald Sull, "Strategy as Simple Rules," *Harvard Business Review* (January 2001).

6. Barry Jaruzelski, Volker Staack, and Brad Goehle, "Proven Paths to Innovation Success," *Strategy + Business*, Winter 2014. (insert 6.12 and the rest of content after figure)

7. Robert Cooper, Scott Edgett and Elko Kleinschmidt, New Product Portfolio Management: Practices and Performance," *Journal of Product Innovation Management* 16, January 1999, pp 333-351.

8. Randal Englund and Robert J. Graham, "From Experience Linking Projects to Strategy," *Journal of Product Innovation Management* 16, January 1999, pp 52-64.

Resolve

Chapter 7
Resolve

Chapter 7 Contents

What to Expect

There are two major initiatives within the Explore & Create segment: resolving the most critical product uncertainties and developing a final product. So far, we have progressed from Investigate (Chapter 5), where we identified, evaluated, and prioritized the assumptions, and Plan (Chapter 6), where the major output was a high-level Product Charter and a detailed Resolve Development Plan.

This chapter will describe how to resolve the most critical uncertainties using the Resolve Loop (Figure 7.1). Each Resolve Loop includes four major iterative steps: **1. Design**, **2. Build**, **3. Execute**, and **4. Learn & Adapt**. These activities are integrated with the ongoing development of the final product, using the Idea Maturity Model (IMM). Be sure to read the Turba case study to find out the surprise ending to the PERS product.

Following the chapter are three appendices that take deeper dives into related topics. Appendix 7A discusses the mechanics of a Resolve Loop. While Appendix 7B defines additional types of prototypes. And Appendix 7C offers another case study on how a team resolved uncertainty for a product called OdorDone.

Figure 7.1: Resolve Loops

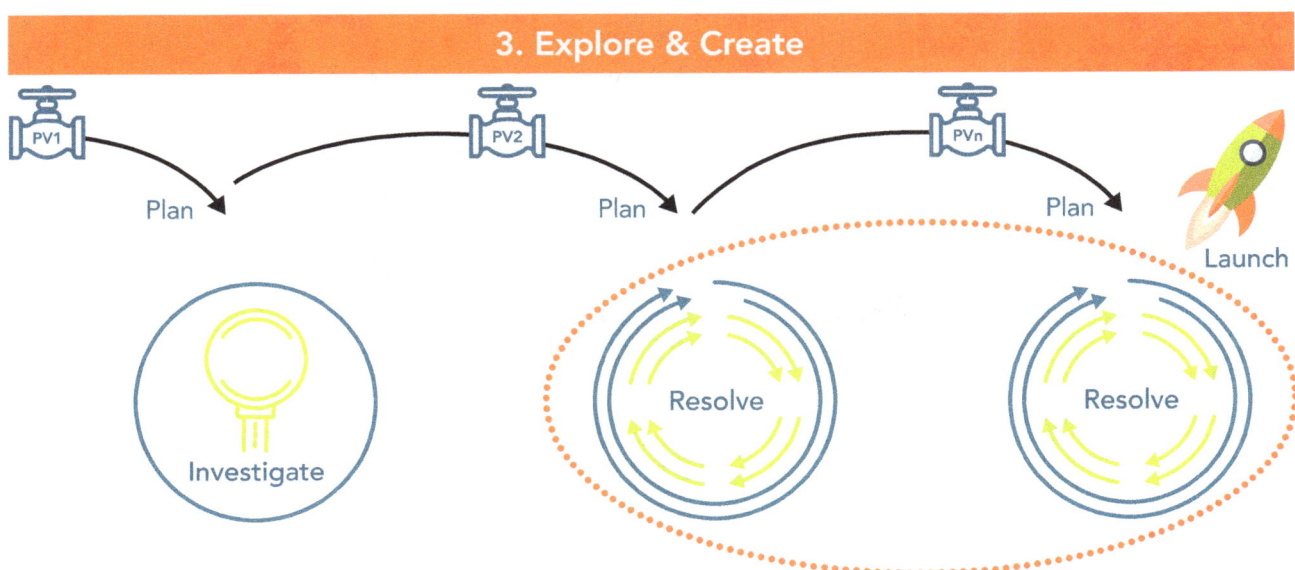

Introduction to Resolve

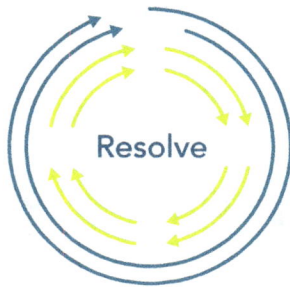

Resolve

Investigate (Chapter 5) and Plan (Chapter 6), prepare the project team to resolve the most critical product uncertainties. Each Resolve Loop addresses an uncertainty that needs to be resolved by the project team. The goal is to reduce the most critical product uncertainties to an acceptable level.

The means of reducing uncertainties can run a gamut from very simple to very complex, from simple data gathering to sophisticated experiments. An excellent example of "simple" is a project where the most significant product uncertainty was getting buy-in from another business unit for resources to deliver on a technology. The VP of engineering resolved the uncertainty within an hour by checking with a counterpart in the other business unit to determine resource availability and interest in co-developing the product.

At the other end of the spectrum, product development for autonomous vehicles (self-driving cars) is highly complex, as discussed in Chapter 1. The number of uncertainties is high due to the complexity of technology and human behavior. Uncertainty lurks in recognition of objects, prediction, systems integration, testing, and willingness to use an autonomous vehicle. According to Bryan Salesky, CEO of Argo AI, these uncertainty areas are complex and hard to solve. As a result, self-driving cars are way off in the future.[1] One of the most significant areas of uncertainty is how much actual demand exists for this technology. The confidence index for self-driving vehicles is 34 out of 100—a very low level of confidence.[2]

Though 2020 was intended to be the year for autonomous vehicles, the industry realized that getting to a "fully automated driving state" would be much more challenging than anticipated. Elon Musk called autonomous vehicles "one of the hardest technical problems that exist—that has maybe ever existed." Also, Uber and Lyft pulled the plug on pursuing autonomous vehicles,[3] and others may follow. Thus, this autonomous vehicle scenario describes uncertainty and risk at an extreme.

When reducing uncertainty, the best outcomes are often achieved through some form of learning, for example, experiments, data gathering and modeling. However, in our experience, some project teams execute learning activities without methodically considering the necessary problem-solving steps. Therefore, we designed a Resolve Loop that integrates four steps: 1. Design, 2. Build, 3. Execute, and 4. Learn & Adapt (Figure 7.2). These activities guide the team in learning.

Four steps to guided learning: Design, Build, Execute, and Learn & Adapt.

Figure 7.2: Four Major Steps of the Resolve Loop

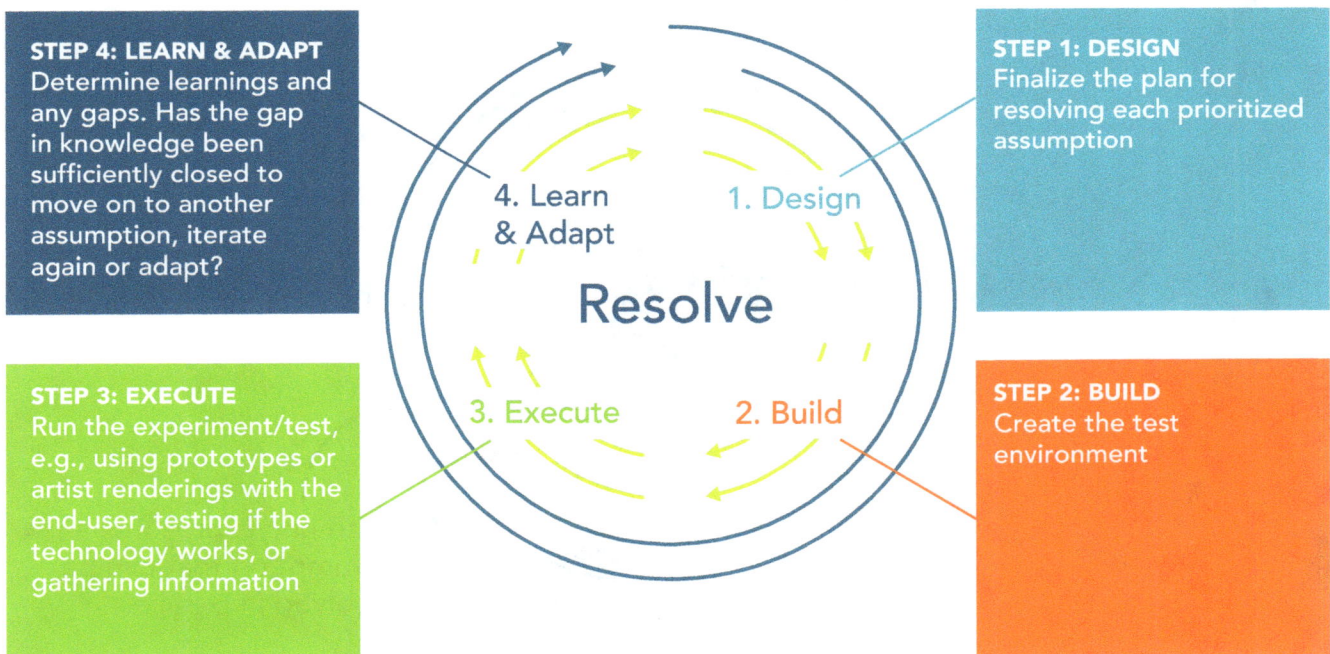

STEP 4: LEARN & ADAPT
Determine learnings and any gaps. Has the gap in knowledge been sufficiently closed to move on to another assumption, iterate again or adapt?

STEP 1: DESIGN
Finalize the plan for resolving each prioritized assumption

STEP 3: EXECUTE
Run the experiment/test, e.g., using prototypes or artist renderings with the end-user, testing if the technology works, or gathering information

STEP 2: BUILD
Create the test environment

4. Learn & Adapt   1. Design

Resolve

3. Execute   2. Build

Similar approaches to experimenting exist. Feel free to follow the steps that best resonate with your organization and team. This chapter provides a high-level overview of each step. Ultimately, the team executes these four steps to understand what they need to learn to reduce product uncertainty.

The Resolve Loops appear to be a series of smooth, perfectly rounded feedback loops, but they can operate more like a maze. The road to uncertainty reduction is not always smooth, and there may be a series of setbacks based on new learnings.[4]

The Resolve Loops appear to be a series of smooth, perfectly rounded feedback loops, but they may operate more like a maze.

As a reminder, we describe uncertainties and risks as assumptions, (Chapter 5, Investigate). Within the Resolve Loop, you can resolve multiple assumptions in parallel (Figure 7.3), but this is only possible if there aren't any dependencies or interrelationships between them. It is rare for just one experiment to resolve all the uncertainty. Although these seem like discrete activities within each step, they can overlap, so there's no need to panic.

Figure 7.3: Multiple Paths to Resolution

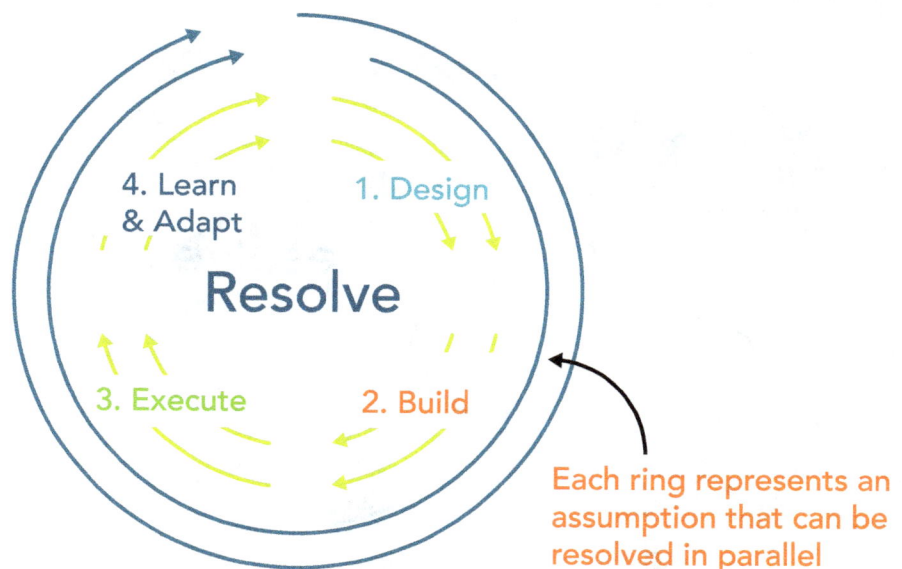

Each ring represents an assumption that can be resolved in parallel

It is rare for just one experiment to resolve all the uncertainty.

Also, as discussed in Sidebar 7.1, you may pursue other approaches to learning besides experimentation.

Sidebar 7.1. Is Experimentation Always Needed?

Experimentation is one method of learning in the quest to resolve assumptions. There are other learning methods available. Data gathering, for example, through secondary research is a relatively quick and inexpensive way to learn from the tests and experiences of others. Various observational techniques, like ethnography, provide valuable first hand information to help understand the uncertainty, and can be used to develop hypotheses for testing.

Learning methods are not very useful when the uncertainty is driven by unknown future outcomes. In these cases, monitoring the situation and devising ways of responding to the future outcome is a way to reduce risk. For example, will a competitor encroach on our marketplace? You can only monitor this, not experiment. This is similar to what the Turba team experienced with competitive threats; they could only monitor the activity of possible PERS entrants (see Sidebar 5.4 in Chapter 5, "Investigate").

Do you want to learn more about the mechanics of the Resolve Loop? If so, please go to Appendix 7A.

Tools of the Trade

Five tools introduced in the previous chapter are carried over into Resolve, including:

1. The **Product Charter** is the high-level document updated throughout the process, aligning with the Resolve Development Plan (RDP) and major project milestones.

2. Only the upcoming Resolve Loops are planned in detail; future loops are roughly scoped. Upon entering a Resolve Loop, you will add more detail to the **Resolve Development Plan (RDP)**. Then, the RDP is updated and built upon at each step in the Resolve Loop (Design, Build, Execute, and Learn & Adapt), as you will see in the Turba RDP examples throughout this chapter (Sidebars 7.2, 7.4, 7.6, and 7.7).

3. The **Pledge** gives the project and management team a solid framework of goals and boundaries. A bilateral agreement between the project team and senior managers specifies the project's budget, resources, objectives, and time.

4. The project team tracks the resolution of the prioritized assumptions using the **Assumptions Tracker** within the Product Risk Framework or a spreadsheet. This helps the team track the resolution of product uncertainty and risk. It also helps identify potential dealbreakers, providing a mechanism to cancel the project before expending valuable resources.

5. The **Idea Maturity Model (IMM)** is an essential tool that guides adaptive planning in ExPD. The project team can record the starting point and the maturity of the idea throughout the ExPD process. As the team adapts the plan to respond to learnings and discoveries, the IMM is the beacon that keeps the team on track toward a viable product.

Get Ready to Resolve Product Uncertainty: The Four Steps

The four steps in a Resolve Loop include the major activities highlighted in Figure 7.4. This section will look at the activities in each step to see how a team resolves the most critical product uncertainties.

The activities described within the four steps are a guide to consider when resolving product uncertainty. Since this is an adaptive process, we recommend executing the activities that make the most sense for your experiment or research.

Also, the four steps in the resolve loop are described here for experimental learning. The same steps can be applied to data gathering and other learning methods. For example, the OdorDone Case Study in Appendix 7C applies the four steps to data gathering instead of experimental learning methods.

Figure 7.4: Major Activities within the Resolve Loop

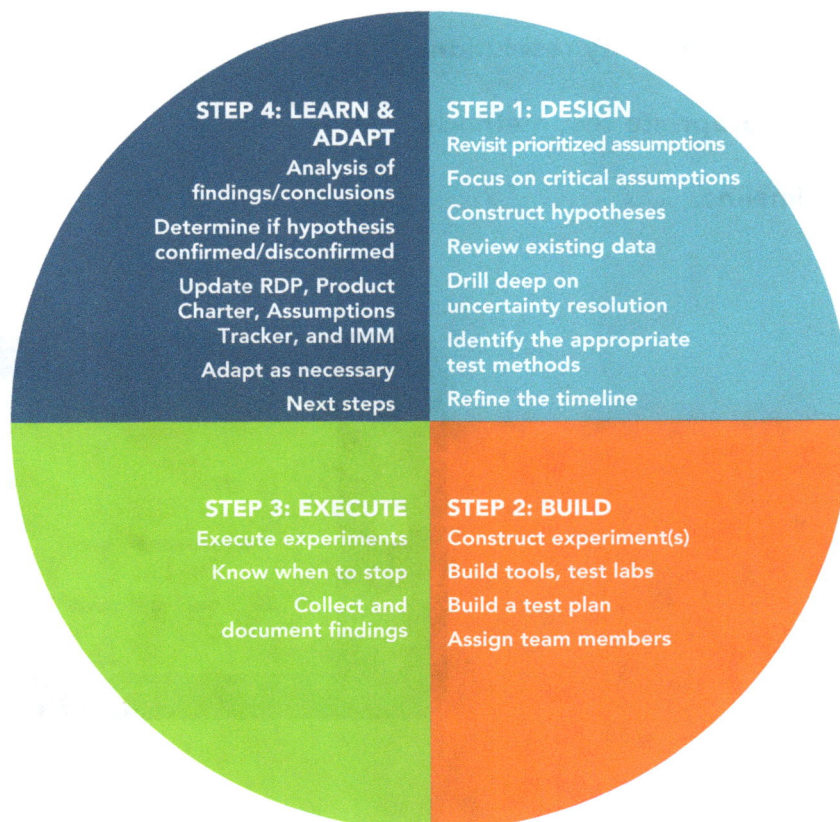

STEP 4: LEARN & ADAPT
Analysis of findings/conclusions
Determine if hypothesis confirmed/disconfirmed
Update RDP, Product Charter, Assumptions Tracker, and IMM
Adapt as necessary
Next steps

STEP 1: DESIGN
Revisit prioritized assumptions
Focus on critical assumptions
Construct hypotheses
Review existing data
Drill deep on uncertainty resolution
Identify the appropriate test methods
Refine the timeline

STEP 3: EXECUTE
Execute experiments
Know when to stop
Collect and document findings

STEP 2: BUILD
Construct experiment(s)
Build tools, test labs
Build a test plan
Assign team members

Design

STEP 1: DESIGN
Finalize the plan for resolving each prioritized assumption

The objective of the 1. Design step is to define the approach and details for how the uncertainty will be resolved. It is front-loaded with more activities than the other three steps. The adage "Fail to plan, plan to fail" is also relevant with experimentation.

The following list of activities will help the project team design the experiment:

1. Revisit prioritized assumptions.

2. Focus on critical assumptions.

3. Construct hypotheses.

4. Review existing data, including past projects and literature reviews.

5. Drill deeper on uncertainty resolution.

6. Identify the appropriate test methods.

7. Refine the timeline.

1. Revisit Prioritized Assumptions

Begin by revisiting prioritized assumptions with the cross-functional team to ensure that the assumptions as stated are still relevant. There may be significant commercial, regulatory, or technical changes, and it is important to revisit any changes. Of course, if the project proceeds and resources are assigned immediately, this step becomes unnecessary. Unfortunately, we have encountered many projects in limbo for months or even years before proceeding in the process.

There may be significant commercial, regulatory, or technical changes, and it is important to revisit any changes.

Confirm the project's objective: What does the team expect to resolve and learn with the most critical assumptions, and what are the goals? Revisit the Product Charter, RDP, Assumptions Tracker, and IMM since things may have changed, particularly if the start of the project was delayed.

2. Focus on Critical Assumptions

After revisiting the assumptions in the previous step, have your team force-rank your most critical assumptions. Then focus your attention on those.

3. Construct Hypotheses

Now convert each assumption into one or more concrete, workable hypotheses that can be tested experimentally.

As you recall from (Sidebar 6.3) Chapter 6, "Plan," the Turba team determined the two most critical assumptions. The project team chose two out of the four hypotheses they created (highlighted in orange) for their initial experiments, starting with the Design step in Table 7.1.

Table 7.1. Example of Creating Test Hypotheses

Turba's Prioritized Assumptions:	Hypotheses:
Target users living at home are willing to pay for a PERS that works everywhere, including outside the home.	• **More than half of seniors want to use the PERS away from home** • Seniors that want the PERS away from home are willing to pay more than $19.99 for a monthly service fee
Bluetooth technology will reliably enable PERS to work anywhere your smartphone works.	• **The Bluetooth technology will perform reliably for the chosen use cases** • The PERS battery life will be at least five days on a full charge

As you iterate, you may refine your hypothesis based on what you find during **4. Learn & Adapt** from existing research, or you may want to understand more about your hypothesis by observing and trying to understand the problem in more detail. The key is to validate your hypothesis through a well-designed experiment.

4. Review Existing Data

Review existing data, including past projects and literature reviews, to glean any learnings that apply to the most critical assumptions. Look at past projects and determine if past designs can be leveraged. This can initially help the team gain a better understanding of the project. You can also work toward constructing a testable hypothesis and think of ways to either confirm or disprove each hypothesis through your experiment.

5. Drill deeper into uncertainty resolution

Now drill deeper into uncertainty resolution. Refine the hypothesis. How will the uncertainty be resolved? What are the tactics, and what are the metrics for success? If possible, come up with several experiments or approaches, analyze them, and select the best approach.

If possible, come up with several experiments or approaches, analyze them, and select the best approach.

Sometimes high risk products require a high level of creativity. Therefore, we recommend that you facilitate a brainstorming session with internal, and if appropriate, external experts who have solid experience in the subject matter of your assumptions. These may be lead users, consultants, academics, human factors engineers, design-thinking experts, and a member of technical staff. During this meeting, you can break complex assumptions into smaller, more manageable components that can be resolved rapidly. This session's outcome typically includes experiments, test methods, needed resources, and metrics that correspond with the project's goals (Figure 7.5).

Figure 7.5: Brainstorming Test Approaches

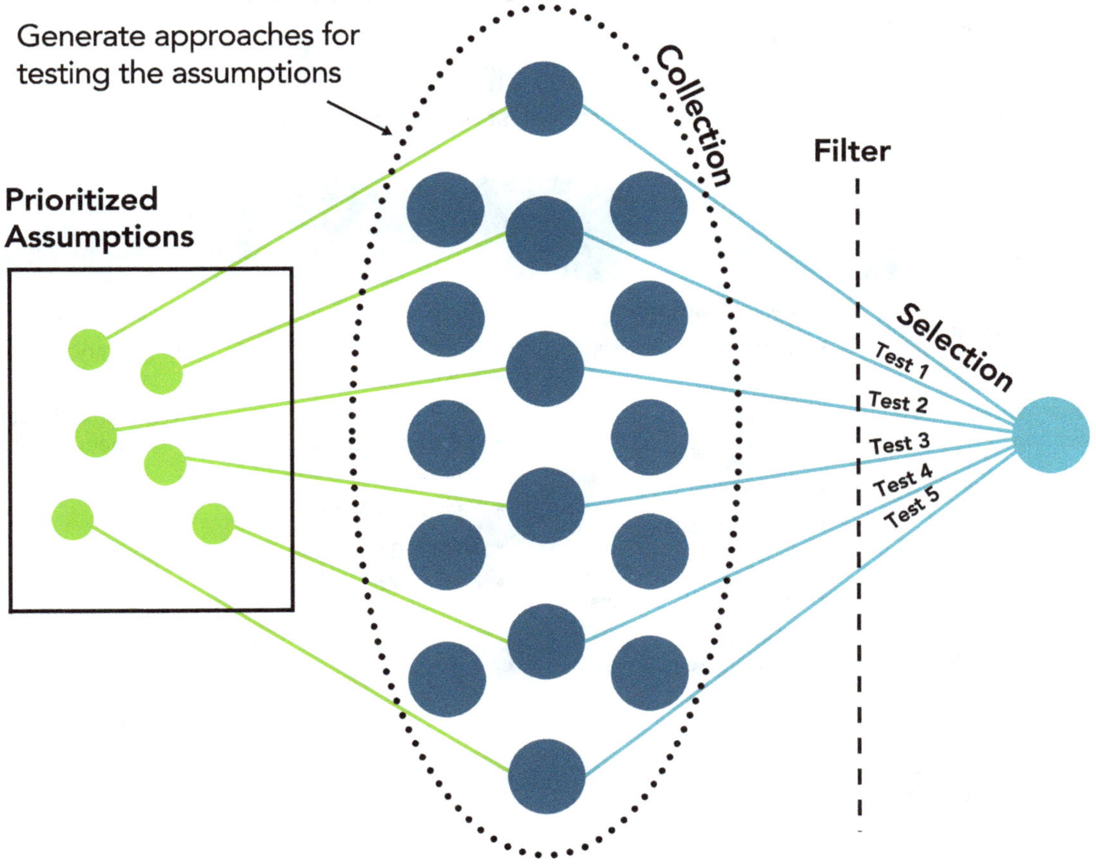

Whomever you invite, make sure that they have something to offer from an alternative perspective. These team members are instrumental in seeing significant gaps in information and identifying any additional uncertainties. Consider extending your invitation to the following disciplines, based on the type of uncertainty you are investigating:

- **Commercial uncertainty.** Consider inviting personnel from finance, upstream and downstream marketing

- **Technical uncertainty.** Include the engineering discipline or scientists who are experts within the product category. Also, invite extended team members, such as lab technicians, regulatory, intellectual-property experts, and operations personnel

6. Identify the appropriate test methods

Use the information gathered so far to identify the most appropriate test methods. Table 7.2 provides thought starters on test methods used across the major technology, commercial, operational, and regulatory categories. We recommend that you choose methods that apply to your project.

Table 7.2. Thought Starters: Test Methods to Assist in Reducing Uncertainty

Technology Tools and Techniques	Commercial Test Methods	Operational Test Methods	Regulatory Test Methods
• Design of Experiments (DOE) • Prototypes • Computer-based optimization • Intellectual-property search • Literature search • Reverse engineering • Breadboard • Test of existing products • User research • Computer simulations • Theoretical (physical or mathematical) modeling • Experimental measurement • Brainstorming • Lab testing	• Market assessment • Competitive analysis • Secondary, primary research • Ethnography • Concept testing • Conjoint testing • Prototyping • Product comparisons • Personas, profiling users • Computer simulation (Monte Carlo analysis) • Modeling • User interviews • Reverse Income Statement	• Supplier capabilities analysis, and research • Manufacturability analysis • Production capacity analysis • Packaging study • Shipping study • Failure Mode Effects Analysis (FMEA)	• Classification research • Country requirements research • Materials investigation to meet regulations

At this time, you will also determine the best test subjects. The test subjects will vary based on the learnings you're trying to uncover. For example, the customer or end-user is the test subject for determining if they will buy. A simulation model may be the targeted test subject to determine if the product is technically feasible.

The test subjects for the prototype will vary based on the findings you're trying to uncover.

As the team constructs the experiments, it must keep in mind these key considerations:

- Keep your cross-functional partners involved

- Call upon outside experts or senior staff for input

- Understand the customer problem and the environment in which the customer will use the product (human factors analysis)

- Understand who will be buying the product and who will be using it, and then determine the difference between the buyer and user

- Understand the cost of experimentation (including materials and resources) and keep it low. If appropriate, use simulation or modeling instead of a physical prototype

These considerations will help the project team focus on critical inputs when constructing an experiment.

Note: Sometimes an experiment can gather data to identify, test, or refine more than one assumption or hypothesis. For example, a Voice of the Customer study can gather information on various user needs, attitudes, and more, that are useful in identifying new assumptions and hypotheses to test. Keep this in mind while designing experiments.

7. Refine the Timeline

Although the work is exploratory, that doesn't mean the team can't follow a timeline. If necessary, refine the RDP's current timeline. It may be challenging to nail down and achieve the timeline, given the Resolve Loops' iterative nature.

As described in Chapter 6, "Plan," start with a high-level view of what it will take to achieve each major milestone within the Product Charter. If product uncertainty is extremely high, you may just start with the scope of effort, cost, and the number of uncertainties to resolve. As the team learns more, milestones can then be integrated.

Ideally, planning should be nimble since these are fast iterations (two to three weeks), and the experiments should be exposing dealbreakers, dead ends, or new discoveries. For a starting point, we recommend that you determine a timeline for each experiment chosen during the brainstorming session, as shown in Figure 7.6.

Figure 7.6: Brainstorming Test Approaches

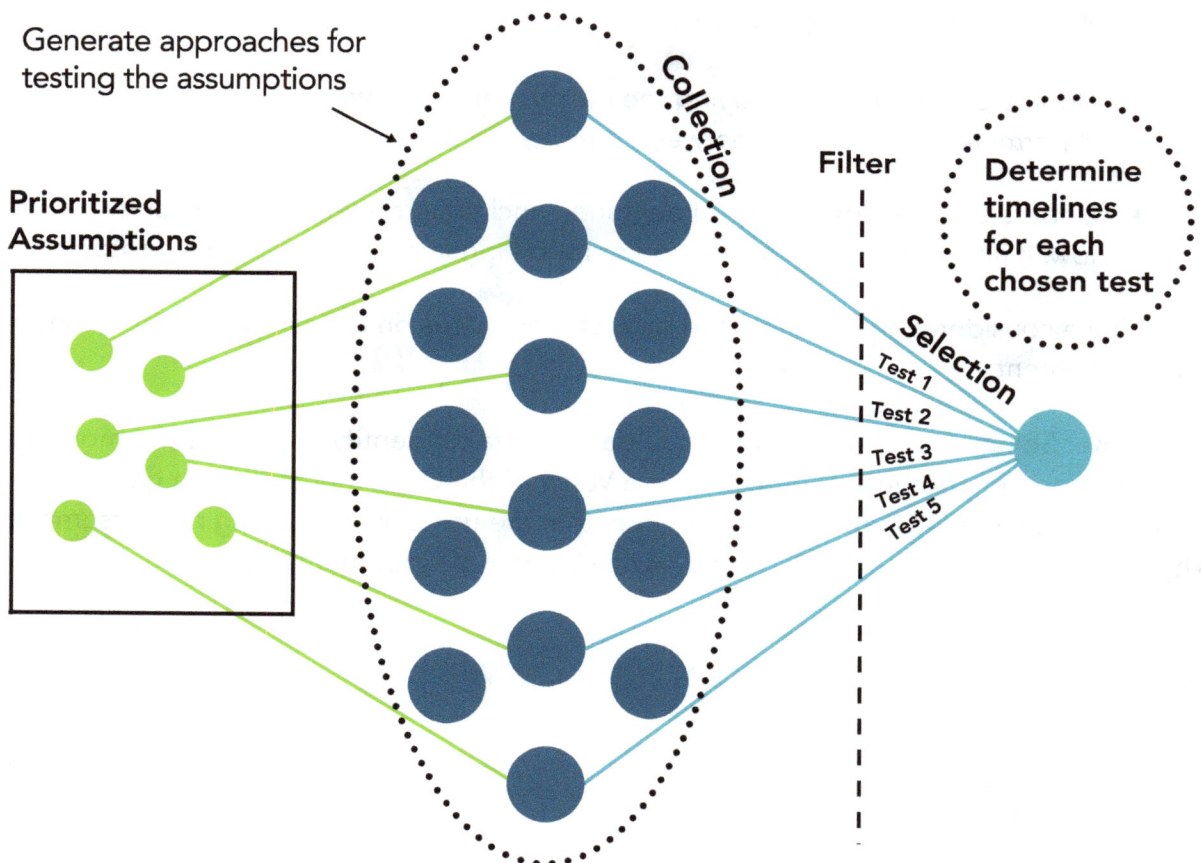

Now that you have an overview of the activities in the Design step, read Sidebar 7.2 to learn how Turba carried out this step.

Sidebar 7.2. Turba's Design Step

TURBA CORPORATION

The Turba project team used the RDP and followed the four steps of the Resolve Loop (Design, Build, Execute, and Learn & Adapt). The team focused on resolving the two most critical assumptions:

STEP 1: DESIGN
Finalize the plan for resolving each prioritized assumption

Hypothesis 1. Target users living at home are willing to pay for a PERS that works everywhere, including outside the home.

Secondary research suggests that this is true with the caregivers, which will help validate that they have a viable value proposition. The proposed PERS system provides the needed protection to seniors outside of the home while providing peace of mind to loved ones and caregivers.

Based on Turba's experience with other product lines, if more than half of target users are interested in the PERS, the market is big enough to be attractive.

Hypothesis 2. The Bluetooth technology will perform reliably for the chosen use cases.

An outsourced research firm conducted a use case study that helped determine the average distance between the PERS and phone inside and outside the user's home. The following use cases were identified:

- Inside: 25 feet/7.62 meters between PERS and phone

- Outside: 33 feet/10 meters between PERS and phone

- 25% of the time, the PERS will be used outside

Since there are no interdependencies between the two assumptions, they can be explored concurrently. Assumption 1 will be headed by product management, and Assumption 2 will be managed by engineering.

The project teams will share findings in real-time via Slack® (internal messaging system) to communicate learnings and if other uncertainties or dealbreakers were discovered. 15-minute stand-up meetings with both owners occur every Friday to ensure integration and major findings. The project manager will provide oversight and integration between the two teams and ensure that they do not go beyond the parameters of the Pledge (Figure 7.7).

Figure 7.7: PERS Pledge

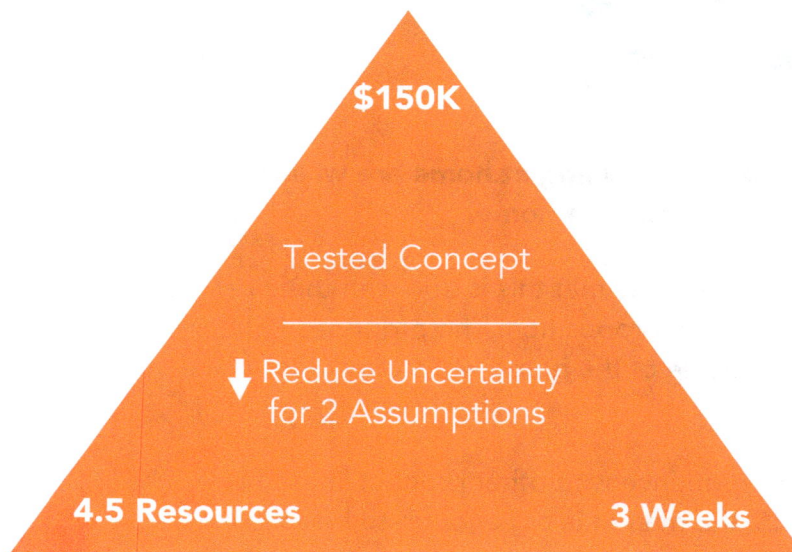

The project manager is also responsible for preparing the detailed project plan to execute the resolution of the two assumptions and the activities needed to reach the Tested Concept (Figure 7.8).

Figure 7.8: PERS Moves from an Idea to a Stated Concept

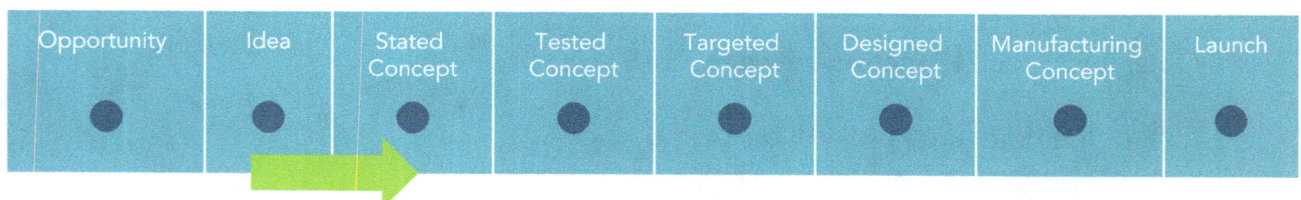

Table 7.3. RDP Example of the Design Step for Turba

	Assumption 1	Assumption 2
ASSUMPTION DESCRIPTION	**Target users living at home are willing to pay for a PERS that works everywhere, including outside the home.** • Based on Turba's experience with other product lines, if more than half of target users are interested in the PERS, the market is big enough to be attractive.	**Bluetooth technology (BT) will reliably enable PERS to work everywhere your smartphone works.** Proposed technology: • Cellular Triangulation and U-TDOA (Uplink-Time Difference of Arrival, the current 911 approved wireless location technology) coupled with two-way voice communication • GPS-based satellite location technology/ Wi-Fi hotspots • Narrowband IoT, LTE-m, and 5G technologies
HYPOTHESES TO TEST	**More than half of seniors want to use the PERS away from home.**	**The Bluetooth technology will perform reliably for the chosen use cases.**
NATURE OF STUDY	**Voice of Customer (VOC) one-on-one interviews with elderly and caregivers.**	**Technical experiments, including but not limited to range studies, effectiveness, packaging capabilities (physical size and weight), environmental resilience (sealing capabilities), power consumption (battery life).**
DEPENDENCIES	N/A	N/A
LEARNING GOALS	**Learning goals:** • Why and how do people use PERS now? • How do they meet that need when away from home? • Is it an important enough problem for the elderly and/or caregivers to pay to solve? • Identify any other relevant hypotheses	**Learning goals:** • Understand the use cases from the research firm: where, why, and how people go outside the home with their phones • What is the minimum acceptable BT Class for the PERS application? • How do use cases impact battery consumption? • Identify any other relevant hypotheses

METRICS	• Important unmet needs identified
	
	 • Level of interest/willingness to pay evaluated
		• BT performance at different distances, inside and outside:
		
		   • Accuracy of data transfer
		
		   • Consistency in maintaining a connection between the PERS and the smartphone
		
		   • Robustness in successfully transferring data at various, even sub-optimal, conditions
		
		   • Battery consumption

OUTPUT	**Test method and timeline** Specify test population and study parameters: • Specify criteria for selecting participants and the number of interviews • Determine whether interviews will be in-person, by phone, or using another approach • Determine how to recruit participants and the appropriate incentives	**Test method and timeline** • DOE to determine best communication/ location protocol • DOE—radio frequency (RF) Antenna geometry Bluetooth chip manufacturing Class of BT
PLEDGE	• Budget: $150,000.00 • Duration: Three weeks • Reduction in uncertainty to an acceptable level • IMM: Currently, at Stated Concept, the objective is to reach Tested Concept • Resources: 1 FTE Product Manager, 3 FTE Engineers, and .5 FTE Project Manager	

PROVIDE A LINK TO A DETAILED SHORT-TERM TIMELINE/PROJECT PLAN FOR THE TWO ASSUMPTIONS AND IMM ACTIVITIES.

Build

STEP 2: BUILD
Create the test environment

The 2. Build step entails creating the test environment and building what you planned in the previous step, 1. Design. At this point in the process, the project team can build a model, survey, conjoint analysis, prototype, or any testing method to execute an experiment.

At this point in the process, the project team can build a model, survey, conjoint analysis, prototype, or any testing method to execute an experiment

Since there are so many different types of product uncertainties, the test methods vary based on the uncertainty you are trying to resolve. As illustrated in Table 7.2, different test methods can be used to reduce technology, commercial, operational, and regulatory uncertainty. Use the test methods and environment that best reduces the product's most critical uncertainties at the lowest cost.

During **2. Build**, the team builds the test methods and environments chosen in **1. Design**. The organization activates resources and the necessary infrastructure (facilities, space, lab, software tools, etc.) to support the test environment. The following list of activities will help the project team build the experiment:

1. Assign team members.

2. Build a test plan.

3. Build tools and test labs.

4. Construct experiments.

1. Assign Team Members

In assigning team members, you must have the appropriate resources (type and number). We have seen some companies assign any available resource, which can be fatal to the project. Ensure that you have the right resources with experience conducting experiments; this is also a good time to have less experienced team members learn in real-time from senior employees.

It is also important that the project stays on track and maintains a cadence. Thomke illustrates the importance of keeping the experiment on track without extreme waiting periods:

"Time passes, other problems creep up, causing distractions, and valuable time has been lost between the cause and effect of experimentation. Long delays contribute to ineffective learning and a lack of urgency. People learn most efficiently when their actions are followed by immediate feedback."[5]

2. Build a Test Plan

Next, build a test plan, sometimes referred to as a test protocol. Table 7.4 is an example of a typical test plan. Most companies have some type of test plan within their organization, so feel free to use yours. This test plan is determined during 2. Build and filled in during testing in 3. Execute.

Table 7.4: Example of a High-level Test Plan

Test Name	Test Method	Hypothesis	Acceptance Criteria including confidence interval assumptions	Samples (include number and description)	Results including Pass/Fail	Next Steps

Building a test plan includes building a data collection method. There are some important elements to data collection, including answers to the following questions:

- How will you collect the data during the experiment? For example, will you use software input or manual counts?

- How can we analyze the data? Can we use statistical software?

- How will the results be presented—graphs, tornado charts, bar charts?

3. Build Tools and Test Labs

You need to build the tools and instruments for measuring the output of your experiment. For example, a survey tool such as SurveyMonkey®, or a sociometric software tool for measuring human behavior, can be used during user research.

Along with tools, you may need to set up test labs. Especially if this is a new product for your organization, you may not have the appropriate lab and tools to develop prototypes or models. If you don't, you can form a partnership with an outside vendor or go through an organization like the American Council of Independent Laboratories (ACIL) to help you locate an appropriate lab for your product.

4. Construct Experiments

The next activity, constructing an experiment to resolve uncertainty, can take many different forms. It can include developing questions for a survey, designing observation forms for ethnography studies, and of course, creating a prototype, which is a very worthwhile test instrument.

We spend more time reviewing prototypes within this step since they have so many important uses during an experiment. Prototypes can be created to help demonstrate how a problem can be solved, ultimately reducing product uncertainty. Also, prototypes can speed up product development because you learn fast and adapt quickly. They do not have to represent the whole design of the product; they can be sections of a design representing a specific value proposition or a critical uncertainty that needs to be resolved.

Using prototypes can speed up product development because you learn fast and adapt quickly.

Prototypes can be used at any point within ExPD. However, we recommend starting early in the Ideas & Selection segment, as described in Chapter 4, "Idea Management System." This is a perfect place to start using low fidelity[6] prototypes or minimum viable products (MVPs)[7] for early customer validation.

To learn about the fidelity of prototypes, refer to Sidebar 7.3.

Sidebar 7.3. The Fidelity of Prototypes

Because of prototypes' different roles, their form and expectations should change as you proceed through the product development process. Thomke refers to the prototype's or model's "fidelity," with low fidelity being the lowest form, a quick-and-dirty prototype. In contrast, high fidelity describes prototypes that typically validate the design and involve high cost and commitment. Low fidelity prototypes are best executed during exploration, and higher fidelity prototypes are used to verify a design.

Prototypes take different forms based on whether they are customer, technical, manufacturing, or testing prototypes. Here we focus on the customer prototype to illustrate the fidelity of prototypes used in various ways.

Prototypes are used to explore customer needs and market acceptance during the earliest parts of product development. Often handmade rough approximations or artist renderings are enough. Don't overthink the prototype early in development. Instead, use something that approximates the concept or a portion of the idea to collect the relevant information. It's the beginning of the conversation with the customer.

For example, you want to understand the acceptable size and weight of a surgical handpiece for use by female surgeons. Hand-carved foam of various form factors and weights, a milled piece of plastic, or an additive manufacturing model may be enough to obtain the desired information. If you are developing software, a paper mock-up of various screens representing the user interface (UI), instead of working software, is often enough to collect information on how the customer might interact with the UI or find value in the concept. Using sticky notes on the mock-up screens allows for real-time modifications and updates with potential users.

As the product is designed, the team should arrange for structured reviews of the product with customers. The objective includes ensuring the product design continues to meet the customers' needs. You may also collect feedback about design trade-offs or information on how a person interacts with the product (human factors engineering). This information provides additional input, which is used to refine your design further.

For more information on the other types of prototypes (technical, manufacturing, and testing), please to Appendix 7B.

Sidebar 7.4. Turba's Build Step

TURBA CORPORATION

STEP 2: BUILD
Create the test environment

Table 7.5 RDP Example of the Build Step for Turba

	Assumption 1	Assumption 2
ASSUMPTION DESCRIPTION	Target users living at home are willing to pay for a PERS that works everywhere, including outside the home.	Bluetooth technology (BT) will reliably enable PERS to work everywhere your smartphone works.
HYPOTHESES TO TEST	More than half of seniors want to use the PERS away from home.	The Bluetooth technology will perform reliably for the chosen use cases.
ACTIVITIES	**1. Develop interview guide:** a) Write the interview guide to obtain the required learnings. b) Test the interview guide on coworkers and refine as necessary. c) Determine which team members will perform the interviews. d) Determine a schedule for the interviews. **2. Determine data collection process:** a) Arrange transcription service. b) Determine how data will be collected and organized. c) Determine the approach for analyzing the data (e.g., affinity diagram).	1. Identify parameters of the test environment. 2. Create test environments (inside and outside). 3. Develop test protocol for chosen metrics. 4. Establish data acquisition and reporting procedures and assign responsibilities.
OUTPUT	• **Interview guide** • **Interview schedule** • **Transcript analysis process**	• **Test plan** • **Accept/reject criteria**

Execute

Once the experiment has been built, the test is executed and data is gathered following your test plan. This approach applies to reducing all types of uncertainty. If the idea is a breakthrough and the task is very exploratory, it is not unusual to iterate back to 1. Design and 2. Build to integrate needed design changes.

The following list of activities will help the project team execute the experiment:

1. Execute the experiment(s) in iterations.

2. Know when to stop.

3. Collect and document findings.

1. Execute the Experiment(s) in Iterations

We recommend that these experiments be executed as quick iterations. The actual time depends on the type of assumption, but typically an iteration of two to three weeks gives the team an initial idea of what it will take to resolve a critical assumption. Structuring a project with short iterations offers benefits, including faster feedback, faster cycle times, and shorter queues.

Structuring a project with short iterations offers benefits, including faster feedback, faster cycle times, and shorter queues.

2. Know When to Stop

We have worked with engineers and scientists who love to tinker and, in some instances, continue to solve the problem until it is 100 percent certain. We have also worked with product managers who do excessive voice-of-the-customer (VOC) research (see Sidebar 7.5). This may result from their personality or reflect an internal culture where failure is not tolerated, and they must be at least 99.9 percent sure.

Sidebar 7.5. When Have I Collected Enough VOC?

Abbie Griffin and John Hauser, in their classic article "The Voice of the Customer,"[8] provide guidance on this question. Based on their data, they stated that "20–30 interviews are necessary to get 90–95% of the customer needs" for a single customer segment.

Before spending millions of dollars on a VOC study, we recommend that you start with a smaller initiative and roll it out over time as you learn more about the uncertainty you're trying to resolve.

Regardless of the motivation to continue, you should know when to stop. But when do you know when you have experimented enough? Unfortunately, there is no absolute answer, especially when you are dealing with a lot of uncertainty.

It is also essential to follow the Pledge in its entirety and not deviate unless there is complete agreement from the team and management. This prevents feature creep or "gold-plating" (overdesigning) from delaying the project.

Part of the Pledge includes timeboxing the experiment. For an illustration, let's say an experiment was timeboxed for 10 working days, and on the 10th day, the team determined that they reduced the uncertainty by 55 percent. Next, the project team and management must determine if uncertainty reduction is sufficient and whether they wish to learn more and continue, adapt, hold, or cancel the project.

3. Collect and Document Findings

The findings are a living document that should be updated regularly once an experiment has been executed. It is best not to wait until completing a project to document the test results. Instead, results should be documented as obtained, whether from an ethnographic research study, a physical design concept study, or a compliance test.

Results should be documented as obtained.

This information can be updated in the test plan, as described in the Build step. Finally, the formal report will summarize the test results and provide a detailed analysis of the collected data as supporting documentation.

To learn the outcome of Turba's **3. Execute** step, see Sidebar 7.6.

Sidebar 7.6. Turba's Execute Step

TURBA CORPORATION

STEP 3: EXECUTE
Run the experiment/test, e.g., using prototypes or artist renderings with the end-user, testing if the technology works, or gathering information

Turba completed the Execute step and updated the RDP, as shown in Table 7.6.

Table 7.6. RDP Example of the Execute Step for Turba

	Assumption 1	Assumption 2
ASSUMPTION DESCRIPTION	Target users living at home are willing to pay for a PERS that works everywhere, including outside the home.	Bluetooth technology (BT) will reliably enable PERS to work everywhere your smartphone works.
HYPOTHESES TO TEST	More than half of seniors want to use the PERS away from home.	The Bluetooth technology will perform reliably for the chosen use cases.
ACTIVITIES	1. Perform interviews. 2. Analyze transcripts. 3. Prepare a summary of findings.	1. Validate test. 2. Perform test. 3. Fill in the test matrix.
OUTPUT	• Summary of findings and targeted areas for next level of research	• Summary of findings and targeted areas for next level of research

Learn & Adapt

STEP 4: LEARN & ADAPT
Determine learnings and any gaps. Has the gap in knowledge been sufficiently closed to move on to another assumption, iterate again or adapt?

It can be challenging to separate 3. Execute from 4. Learn & Adapt since these two steps are closely intertwined. During 4. Learn & Adapt , assimilate what has been learned, draw conclusions, determine major gaps, adapt as necessary, and define next steps.

The following list of activities will help the project team learn from the experiment:

1. Analyze findings and draw conclusions.

2. Determine if the hypothesis is confirmed or disproved.

3. Decide the next steps.

1. Analyze Findings and Draw Conclusions

After executing the experiment, we recommend that the team review the results to determine if they have filled the gaps in their knowledge. What was learned? Do we still have gaps? Were new uncertainties discovered during the experiment? If so, add these new uncertainties to the Assumptions Tracker so that the team can decide on the next steps.

During analysis, it is important to draw conclusions from the experiment. During each step, the researcher needs to ask, "Do these results make sense?" It is also important to verify findings. Verification can be achieved by replicating runs or, in some cases, the entire experiment.[9]

We recommend that design reviews involve engineers who do not directly work on the project to offer valuable feedback on experiments. Also, these engineers can be involved in reviews during the experimentation process (Design, Build, Execute, and Learn & Adapt).

We recommend that design reviews involve engineers who do not directly work on the project; they can offer valuable feedback on experiments.

2. Determine if the Hypothesis Is Confirmed or Disproved

Next, assess the research results in terms of the hypothesis. Did your research support and confirm the hypothesis? Or did it disprove your original hypothesis? Regardless of the outcome, learnings are of paramount importance. Hopefully, the research provided more insight and certainty on the assumption.

3. Decide the Next Steps

Based on the learning and gaps, the team needs to decide on the next steps. One of the critical questions is, "Did we resolve enough uncertainty?" It may be advisable to put together a review committee (management, senior fellow, and project team) to provide input on the results, especially if the research involved a high-cost experiment.

The Resolve Loop offers four routes you can take as your next steps:

A. **Continue and learn more.** You can continue to iterate through the four steps of the Resolve Loop, starting with **1. Design**. This can involve either the original experiment or revised assumptions and hypotheses.

B. **Put the project on hold.** Putting a project on hold can happen for many reasons, for example, a lack of resources or budget. It is crucial when a project is restored, the assumptions should be revisited since they have probably changed.

C. **Adapt.** Adapting can be the best option when you learn something new that indicates a better way forward. For example, during experimentation, the team could discover an alternative market for the product, discover new technology, or an entirely new product that can better meet customers' needs.

D. **Cancel the project.** Reasons for canceling a project may include too much uncertainty and risk, a market being too small, no hope of meeting customer requirements, or technology not being feasible. Also, a potential unforeseen dealbreaker may have surfaced. It is essential to adapt to these changes quickly. For example, suppose a competitor has already launched a product that solves the customer's problem at a lower cost. In that case, the team will need to make a hard decision on whether it makes sense to continue with the project.

We have found that deciding to cancel a project is one of the most difficult decisions for employees, typically due to the significant money invested in the project and the fear of making a mistake. This is another place where the ExPD process can help. Big-spending should not occur until the most critical product uncertainties have been resolved. Making decisions to cancel a project should be easier earlier in the product development process when less skin is in the game.

We have found that deciding to cancel a project is one of the most difficult decisions for employees to make.

During Learn & Adapt, the team also updates the Product Charter, RDP, Pledge, Assumptions Tracker, and IMM.

Read Sidebar 7.7 to understand what the team learned and find out what happened to the PERS product.

Sidebar 7.7. Turba Learns and Adapts Making a Big Decision on the PERS Product

TURBA CORPORATION

STEP 4: LEARN & ADAPT
Determine learnings and any gaps. Has the gap in knowledge been sufficiently closed to move on to another assumption, iterate again or adapt?

Table 7.7. RDP Example of the Learn & Adapt Step for Turba

		Assumption 1	Assumption 2
ASSUMPTION DESCRIPTION		Target users living at home are willing to pay for a PERS that works everywhere, including outside the home.	Bluetooth technology (BT) will reliably enable PERS to work everywhere your smartphone works.
HYPOTHESES TO TEST		More than half of seniors want to use the PERS away from home.	The Bluetooth technology will perform reliably for the chosen use cases.
EVALUATE HYPOTHESES		1. Do we have sufficient qualitative evidence to justify the hypothesis and invest in a self-stated importance survey?	1. Analyze data and describe use cases of the PERS. 2. Decision on whether the hypothesis is supported.
EVALUATE LEARNING GOALS (Answer the learning goals identified in the Design step)		**Learning goals were met:** a. Identified the important unmet needs/problems. b. Indication on willingness to pay for a PERS that works everywhere.	**Learning goals were met:** a. Acceptable Bluetooth performance in identified use cases. b. BT Class 2 is the chosen class for PERS based on the findings.
DETERMINE RECOMMENDATION ON NEXT STEPS		• Continue to self-stated importance survey • Draft a high-level plan for the next step, the self-stated importance survey	• Test for the next hypothesis: The PERS battery life will be at least five days on a full charge
IDEA MATURITY MODEL (IMM)		Turba advanced to a Tested Concept	
VISION		The Turba team felt confident that they were still aligned with the product vision (Sidebar 3.3. Chapter 3, "Why Strategy Matters for Product Development").	
NEXT STEPS		Recommendation on next steps: The Turba team decided to continue with the project. **Next steps:** • Include an updated RDP and detailed plan for the next iteration with a revised Pledge • Update the Product Charter reflecting the project updates with a corresponding high-level plan • Update the Assumptions Tracker	

In evaluating the hypotheses and learnings, a typical question for the Turba team to ask is whether the assumption or hypothesis has been sufficiently resolved. In this case, with Assumption 1, a qualitative study on a small sample via interviews helped the team better understand the nature of the opportunity and get an early indication of whether the hypothesis may be true. However, a small sample is insufficient to evaluate the hypothesis confidently. This would require a second iteration with a larger sample and a better understanding of the opportunity using the self-stated importance survey. This will also help inform the proposed customer segments.

With Assumption 2, the engineers confirmed their hypothesis. The use cases were validated and refined, and they determined that Bluetooth Class 2 was the best choice for the Turba application since it had the lowest battery consumption compared to Class 3.

In completing this work, the Turba project team also progressed the PERS project from a Stated Concept to a Tested Concept in the Idea Maturity Model.

The management team was especially pleased that the team was able to stay within the parameters of the Pledge. They successfully met their objectives of reaching the Tested Concept and reducing uncertainty on the two most critical assumptions. They achieved this in three weeks with resources of four and a half FTEs and a budget of less than $150,000.00. There was a reason to celebrate, and a special luncheon was held for the project and management team.

When the product manager returned from the celebratory project lunch, she noticed a Google Alert on a competitor called JinQ. She monitored possible entrants since it was identified as a critical uncertainty during Investigate in the External Factors Pod (see Sidebar 5.4 in Chapter 5, Investigate).

JinQ, a start-up, announced its introduction of a PERS product very similar to Turba's. However, the JinQ product had better features and functionality, and its price was much lower than Turba could match. It was to be showcased at the next Consumer Electronics Show (CES) in 2022.

The product manager broke the news to both the project and management teams. They deliberated for a while that afternoon and decided to sleep on it. The following day, they reconvened to discuss the facts, data, and the pros and cons of the project, taking a Systems 2 approach (Sidebar 4.5 in Chapter 4, Idea Management System) to making a decision. They realized that they had to adapt quickly before spending more money and important resources on a project that might never catch up to the JinQ product, much less exceed it.

Luckily, the team had put the time into the upfront planning of the product strategies, business model, platform strategy, and roadmap, and they could adapt quickly. They ultimately decided to pivot from the Basic model to the Max model to displace the JinQ product. The sensor technology is currently available to support the Max model. Also, the project team had covered enough groundwork and had the knowledge to develop the Max model for the 2023 CES show successfully.

The team immediately started the Max model project that afternoon, identifying, evaluating, and prioritizing the most critical uncertainties.

The Turba examples throughout this book demonstrate how the ExPD process works. It is a story of product adaptability that responded quickly to competitive uncertainty. The ExPD process gave the Turba team enough structure but not too much. The practice of adapting a project based on a sound strategy was quickly applied before further resources and budgets were spent.

You can read another ExPD case study in Appendix 7C. This case study takes the reader through the ExPD process from Investigate to Resolve for a product called OdorDone.

The next chapter (People) explores the enterprise, management, teams, and culture that supports an adaptive product development process like ExPD.

Key Chapter Points

1. With complex uncertainties, the best outcomes are often achieved through experimentation. For this, ExPD includes Resolve Loops that integrate four iterative steps: 1. Design, 2. Build, 3. Execute, and 4. Learn & Adapt.

2. The four step process can be applied to other learning methods, such as data gathering and modeling.

3. The objective of the 1. Design step is to define the approach and details for how the uncertainty will be resolved. Therefore, the Design step is front-loaded with more activities than the other three steps.

4. The 2. Build step entails creating the test environment and building what you planned in the previous step, 1. Design.

5. Once the experiment has been built, the test is executed and data is gathered following the test plan.

6. During 4. Learn & Adapt, assimilate what has been learned, draw conclusions, determine major gaps, adapt as necessary, and plan next steps.

7. It may be advisable to put together a review committee to provide feedback and input on the results, especially for an experiment that involves significant cost.

Appendix 7A: The Mechanics of the Resolve Loop

This appendix provides an overview of the Resolve Loop's mechanics (Figure 7A.1).

Figure 7A.1: The Mechanics of the Resolve Loop

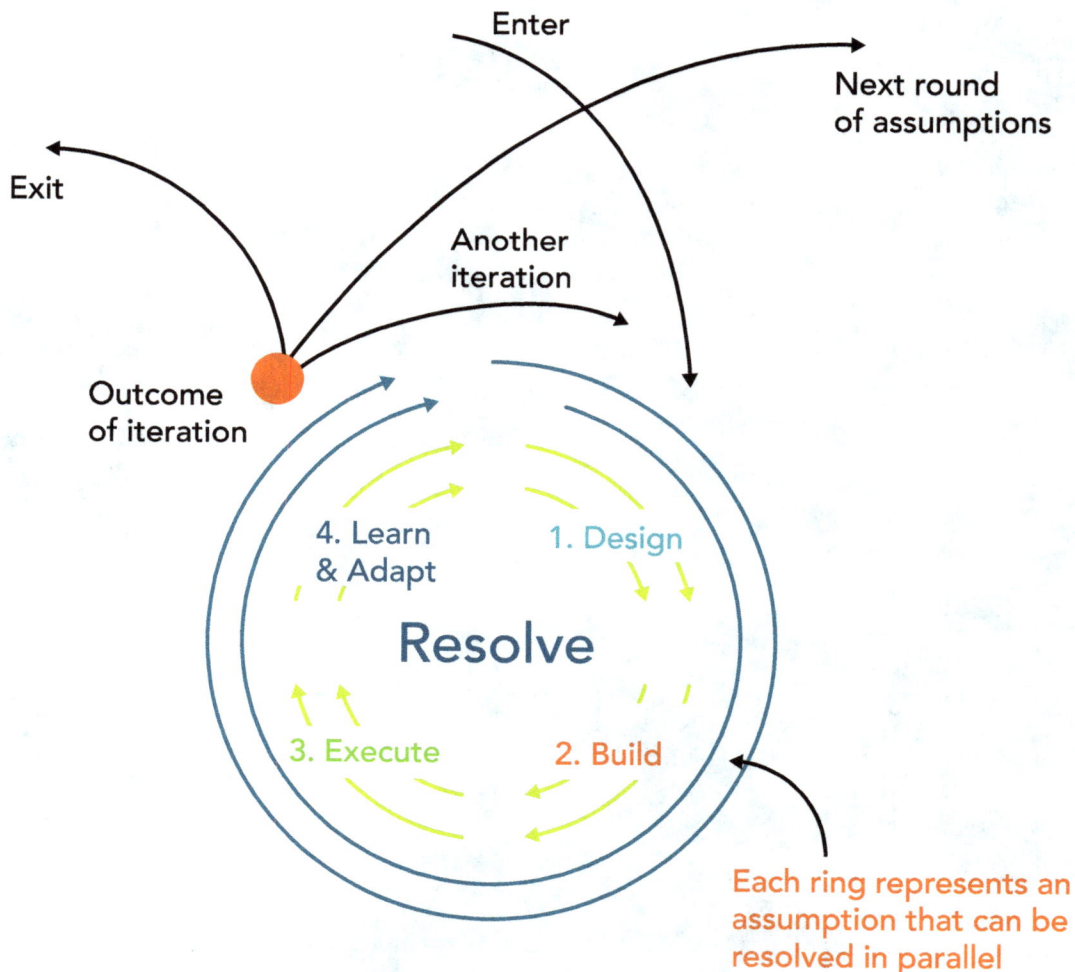

Enter

Next round of assumptions

Exit

Another iteration

Outcome of iteration

4. Learn & Adapt

1. Design

Resolve

3. Execute

2. Build

Each ring represents an assumption that can be resolved in parallel

Enter

You enter the loop with your critical prioritized assumptions. Then you engage in the Design, Build, Execute, and Learn & Adapt activities.

The Rings

Each ring of the Resolve Loop represents an assumption. Depending on the project, you can run assumptions in parallel if there are no interdependencies between them. It is rare for just one assumption or experiment to resolve all the uncertainty, so it is best to conduct multiple small-scale experiments in parallel, especially early in the project.

Another Iteration

Sometimes the team decides that they don't have enough information to resolve the assumption. In that case, another iteration (Design, Build, Execute, and Learn & Adapt) is needed. This decision is labeled "Another Iteration" at the top of the figure.

Next Round of Assumptions

At the upper right of the figure is the path called "Next Round of Assumptions." Follow this path when you have additional prioritized assumptions to resolve.

Exit

You can exit the loop (upper left in Figure 7A.1) in several situations:

- Uncertainty is resolved

- Uncertainty cannot be reduced to an acceptable level, so the team cancels the project

- The team decides to pivot the product in another direction

- The experiment or project needs to be placed on hold due to lack of resources, bad market timing, etc.

Appendix 7B: Categories of Prototypes

Along with customer prototypes, discussed in Sidebar 7.3, an organization may use technical prototypes, manufacturing prototypes, and testing prototypes for various aspects of product development.

Technical

Technical prototypes are sometimes referred to as engineering prototypes. Some technology can take years to develop, and you may need to try many different prototypes before you have something that works. Early in product development, the organization must determine if its current technical expertise can turn the idea into a final product. The expertise needed may be related to the entire product or only a component. You can then use prototypes to start building learnings based on the identified gaps.

Like customer prototypes, technical prototypes often start as rough approximations refined throughout development. It is important to include product management in developing these prototypes to ensure alignment with customer needs. Also, a prototype used to determine technical feasibility may help test market acceptance.

Manufacturing

Manufacturing prototypes are sometimes referred to as production prototypes. The manufacturing prototype should closely reflect the actual product deliverable. It is important to include your manufacturing organization early in the product development process before the design is finalized.

Prototyping also facilitates deciding whether to make the product (manufacture internally) or buy the product or components from an outside source (partner). Using a pilot line to assemble components can build knowledge within manufacturing. The organization can also apply lean principles to the manufacturing process.

Testing

Subassembly and final-product testing can be completed using prototypes. First, however, you will need to document how the prototype represents the final product and what confirmation or new testing will be required to show the prototype data is acceptable. For example, suppose your prototype used a 2-cavity mold, but the production product will use a 12-cavity mold. If the materials are the same, any material-compatibility studies probably can be used, but you probably need to recheck molded-product shrinkage, cooling, and stress points.

Appendix 7C: ExPD Case Study—OdorDone

This case study illustrates how ExPD worked in an organization. It is based on our experiences with actual companies, but the details are disguised to preserve confidentiality.

A medical device organization had a product line of external incontinence devices, and a new product idea came from one of the sales reps. The rep had observed a nurse wrapping garbage bags around the tubing between the patient and the collection bag. The nurse was attempting to reduce the odor escaping through the material of the tubing.

The organization had already done a lot of work to improve the length of time the collection bag would contain the odor before it permeated the material and escaped into the air. They had introduced a new collection bag, which received rave technical reviews, but the sales rep saw that at least one nurse felt the tubing was still a significant odor source. The bag alone didn't solve the problem. The sales rep, with the support of the rest of the project team (product management, R&D, and marketing), suggested the organization create OdorDone tubing. The management committee recognized the significant potential in the idea and gave it a top priority in PV1. The idea passed PV1 and moved to Investigate.

During the initial Investigate project meeting, the cross-functional team determined a clear statement of benefit and need, supported by the research done on an earlier project. The idea was at the 'Idea' level of the Idea Maturity Model.

Using the Product Risk Framework (PRF) to prompt them through the major knowns and unknowns, the team determined that everything listed in Figure 7C.1 in the **black roman (upright)** type was known or fixed. For example, the business model was well established and fixed; it acted as a constraint. The sales channels were already in place and couldn't be changed. The organization already sold the collection bags and tubing, and they knew that the product idea aligned with their strategy. They understood the customer, the size of the market, and unit sales. They also knew a lot about the product and how to manufacture it.

Figure 7C.1. Uncertainty Analysis for OdorDone Tubing

OdorDone Tubing – Product Risk Framework™ (PRF)

Identify What Is Known/Fixed vs. What Is Unknown/Uncertain:

Business Configuration	Product Definition	Strategic Fit	Risk	Financials
Supply	Intended market user	Business strategy	Familiarity with product	Unit sales
Research/ Development/ Engineering	*Problem perceived*	Product, innovation, platform, market, technology strategies	Familiarity with market	*Unit margin*
Manufacturing	*Confimed need*		*Familiarity with technology*	*Cost per unit*
Sales/ Marketing	*Benefits*			*Sale price per unit*
Distribution	*Value proposition*			*Development cost*
Service	*Technology*			Launch cost
	Mandatory features/ performance			

Black – Known/fixed   *Orange/italics – Hypothesized/undefined*

Everything in *orange (italics)* was deemed unknown or uncertain and was prioritized. The most significant uncertainties centered on the specifics of the odor elimination requirements and the materials or technologies used. The team captured this information in the Assumptions Tracker for future updating and modifying. In addition, they prioritized two critical uncertainties, expressing each in the form of an assumption that must be true for the idea to succeed:

1. We can find material appropriate for tubing that will also contain odor.

2. Hospitals, nursing homes, and other customers will be willing to pay extra for this odor reduction.

The team submitted a Product Charter and Resolve Development Plan (RDP) to the management committee with the approach for resolving the uncertainties. During PV2, the committee assessed and prioritized the project, and they concluded that the uncertainties were not insurmountable, so they decided to proceed with the project. The appropriate resources were available to resolve the two assumptions, and PV2 indicated that the project could enter the Resolve Loop in two weeks.

For the first Resolve Loop, engineering and product management worked in parallel on resolving these assumptions, since there were no interdependencies. Let's look at each assumption in greater detail.

Assumption 1: We Can Find an Appropriate Odor-Containing Material

Figure 7C.2: Three Iterations

OdorDone Tubing Resolve Loop 1

Iteration 1: Assess current inventory of technologies on suitability for tubing and the ability to block odor

Iteration 2: Evaluate materials of competitive and related products on ability to block odor

Iteration 3: Licensing opportunities

continue

continue

1.4. Learn & Adapt: Document learnings, update assumptions

1.1. Design: Review current technologies and IP

2.4. Learn: Document learnings, update assumptions

2.1. Design: Review competitive/related technologies and patents

Resolve

Resolve

1.3. Execute: Gather information and analyze

1.2. Build: Identify sources of info

2.3. Execute: Gather information and analyze

2.2. Build: Identify competitor and related products

Engineering's objective was to find a material to contain odor. The engineering team identified three iterations for their analysis (Figure 7C.2):

1. Assess the current inventory of technologies to determine their suitability for tubing and ability to block odor.

2. Evaluate the materials of competitive and related products to determine their ability to block odor.

3. Research technologies for licensing opportunities.

Following the cycle of Design, Build, Execute, and Learn & Adapt (Figure 7C.3), the team evaluated the current inventory of technologies:

Figure 7C.3: Four Major Steps of the Resolve Loop

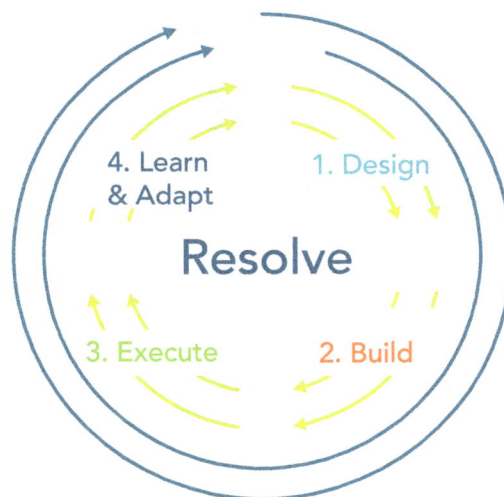

- **Design** involved identifying what information was needed—for example, what criteria to look for in candidate technologies and how to rate them. In addition, the team looked for technologies with a long permeation time

- In the **Build** step, they identified where to look for the information, including intellectual property databases. They also identified experts to interview

- **Execute** involved accessing the databases and interviewing experts. From there, the team analyzed and rated each technology on the criteria identified in the Design step

- In the **Learn & Adapt** step, they concluded from the analysis whether any candidate technologies were available in-house

In the first iteration of this cycle, the team found two possible technologies internally but decided to keep looking. They decided it was too risky to rely on only two candidate technologies after a cursory evaluation, especially since the material was vital to the product's success. Being wrong on this assumption would have a significant impact.

In the second iteration, they did essentially the same thing, looking externally at the materials used by competitors and related products. Ultimately, they found only one material that might be a good fit.

They went to a third iteration, contacting organizations developing new materials to see what they had developed. They found a promising material that reduced the odor significantly, and they decided to license this technology. The team felt confident they had reduced the technology uncertainty enough to launch the product successfully.

Assumption 2: Customers Will Be Willing to Pay for Odor Reduction

Concurrently, the product manager and the project team resolved the second assumption. They reviewed recent customer studies for evidence of the importance of odor containment. They also identified differences by various segments of caregivers, purchasers, and end-users.

Figure 7C.4: Resolving Assumption 2: Willingness to Pay

Iteration 1: Reviewing recent customer studies for evidence of the importance of odor control

1.4. Learn & Adapt: Document learnings, update assumptions

1.1. Design: Review past customer studies for evidence of the importance of odor control

Resolve

1.3. Execute: Gather information and analyze

1.2. Build: Identify sources of info

After just the first iteration (Figure 7C.4), the team was sure they had a winner. A study conducted a year earlier saw that odor was almost always at the top of the list of problems mentioned by caregivers, and there was great dissatisfaction with the current solution. The product manager also knew from previous customer studies that various customer segments were willing to pay to reduce the odor significantly. Based on these learnings, the team hypothesized that there would be a demand for odor-controlling tubing if it provided a significant odor reduction. The team felt satisfied with their findings that the commercial uncertainty was resolved.

Management was pleased with the outcomes and verified that enough resources were available to continue the project until launch.

Interpreting the Case Study

This project team had previously used the phased-and-gated process, and this was the first time they used the ExPD process. Instead of executing the phased-and-gated prescribed activities, the team worked cross-functionally and concurrently and adapted to the product needs versus process needs. They pulled the technology uncertainty to the earliest point in the project instead of executing the task of reducing technology uncertainty in the development phase, as is typical of the phased-and-gated process (Figure 7C.5). Also, the team didn't waste time executing other early-stage activities, like forecasting unit sales in the scoping and business case phase, because they already had evidence that there was a market.

Figure 7C.5: Prescribed Activities in the Phased-and-Gated Process

| 0 Discovery | 1 | 1 Scoping | 2 | 2 Business Case | 3 | 3 Development | 4 | 4 Verification & Validation | 5 | 5 Launch |

The team accomplished two major objectives by moving technology and commercial uncertainty earlier in the process. First, they made it possible to kill the project quickly if the technology failed or if the market was unattractive. Second, they reduced technology- and commercial-related delays. Other benefits included speed, adaptability, strategic alignment, risk reduction, real-time project prioritization, resource optimization, and decreased bureaucracy and paperwork.

Notes

1. Neal E. Boudette, "Despite High Hopes, Self-Driving Cars Are 'Way in the Future," *New York Times*, July 17, 2019.

2. Alex Eule, "Are We There Yet? Self-Driving Has a Long Road Ahead," *Barron's*, May 3, 2021.

3. Ibid.

4. Melisa Buie, *Problem Solving for New Engineers: What Every Engineering Manager Wants You to Know* (Boca Raton, FL: CRC Press, 2018).

5. Stefan H. Thomke, *Experimentation Matters* (Boston: Harvard Business School Press, 2003).

6. Ibid.

7. Steve Blank, *The Startup Owner's Manual: The Step-by-Step Guide for Building a Great Company* (Pescadero, CA: K&S Ranch Press, March 2012).

8. Abbie Griffin and John R. Hauser, "The Voice of the Customer," *Marketing Science* 12(1) (Winter 1993).

9. Buie, *Problem Solving for New Engineers*.

Part III

Adaptive Practices that Support ExPD

Chapter 8
People

Chapter 8 Contents

What to Expect

ExPD success requires the adoption of a new mindset for product developers, and the role of people should not be underestimated. In any organization, people shape an **enterprise**, comprise **management**, structure **teams**, and form a **culture** (Figure 8.1). Each of these roles will be explored further in this chapter.

Figure 8.1: Interrelationship of People

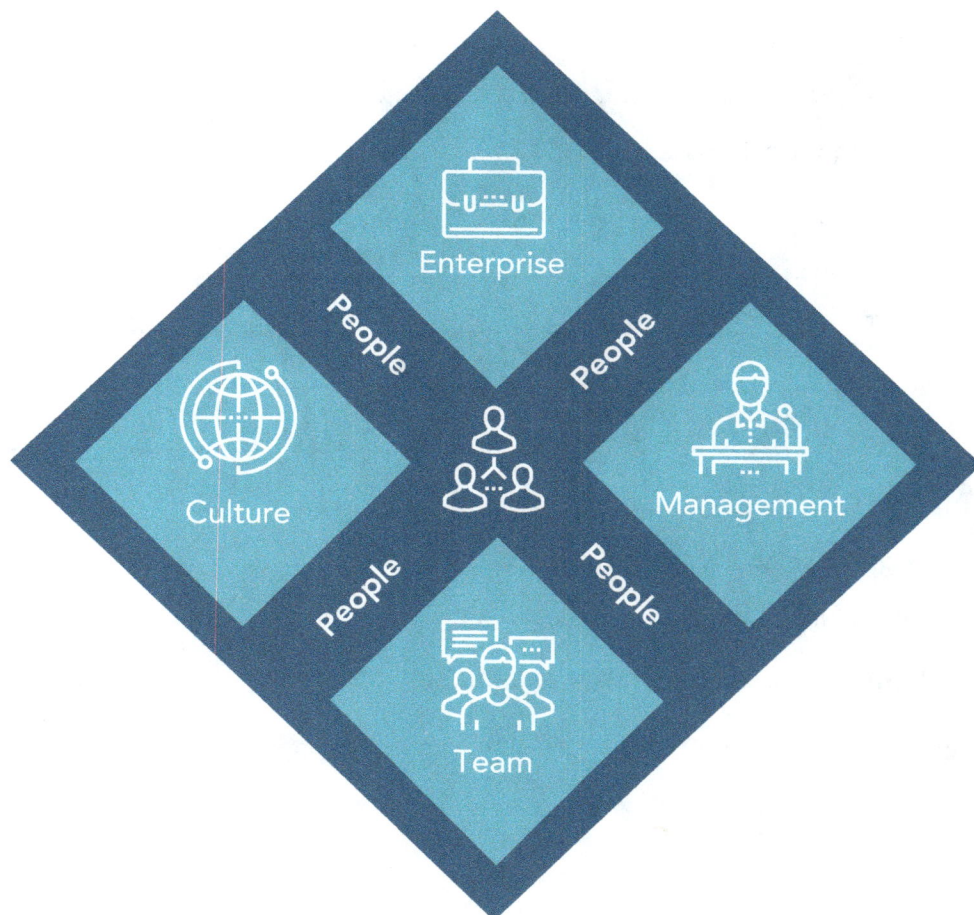

People shape an organization, comprise management, structure teams, and form a culture.

We begin this chapter with an overview of how **enterprises** have evolved to the current state of increased complexity and how speed and adaptability have become critical for survival in today's environment. We will also review whether it is possible for an enterprise to be entrepreneurial, along with the pros and cons of internal incubators and the importance of fostering a learning organization.

This chapter also addresses the critical role of **management** in effective decision-making and intentional change. We will review management methods that support ExPD, including decentralized decision-making and management by exception. We will also draw from important military techniques, including mission-type orders and the entrepreneurial approach of leading by asking questions.

In the **team** section, we discuss the importance of cross-functional partners and building trust. We conclude this section by considering the appropriate team members that comprise an ExPD team.

Finally, we look at **culture**, which is often invisible yet affects all interactions and represents one of the most significant challenges. We discuss it last because culture tends to exist at the intersection of management, teams, and processes. We will address issues related to culture: risk-taking, mistakes, and the stigma of failure.

In each area, we will review practices and methodologies that enable adaptability for ExPD (Figure 8.2).

Figure 8.2: Adaptive Methods

ADAPTABILITY

Techniques
Methodologies
Principles

ENTERPRISE

Internal
Incubators

Learning
Organization

MANAGEMENT

Decentralized
Decision-making

Management by
Exception

Leading by
Asking Questions

CULTURE

Culture
Triangle

Learn
from Failure

TEAM

Cross-
Collaboration

Team Trust

Employee
Mastery

Enterprise

Recent trends in business strategy indicate a shift from the creation and protection of competitive advantage towards survival and adaptation to changes to the external environment.

Table 8.1 highlights the elements of uncertain environments versus the more predictable ones typical in decades past when large enterprises survived from year to year with less need for internal processes, management styles, and culture changes.[1]

Table 8.1. Elements of Predictable and Uncertain Environments

	Predictable Environment	Uncertain Environment
MANTRA	"Be efficient."	"Be adaptive."
STRATEGY	Revisit once a year	Adapt as needed to change in the environment (e.g., commercial, regulatory, and technology)
PROCESS	Prescribed activities (phased-and-gated approach)	ExPD
PROBLEM-SOLVING METHOD	Follow a process	Experiment
MANAGEMENT STYLE	Command and control	Decentralized decision-making
WORKING ENVIRONMENT	Continuous feedback from management	Parameters for work (Pledge)
CULTURE	Fear of making mistakes	Learn and adapt

The differences between the predictable and uncertain environments are evident. In an uncertain environment, complexity is increased, requiring an entirely different mentality and set of practices. Many traditional enterprises are struggling since the new mentality is a significant departure from past practices in a safe and predictable environment.

Startup vs. Enterprise:

In seeking a path towards greater adaptability, some enterprises strive to adopt the practices of startups since they are the epitome of adaptability. Steve Blank, a serial entrepreneur and founder of the Lean Startup movement, describes the difference between an established enterprise and a startup:

"An enterprise is a permanent organization designed to execute a repeatable and scalable business model, while a startup is a temporary organization designed to search for a repeatable and scalable business model."[2]

As mentioned in Chapter 1, "The Case for ExPD," the Lean Startup movement grew mainly from the entrepreneurial internet-based software community, where startups looked for venture capital backing. The Lean Startup approach focuses on creating a working business model and finding an appropriate product-market fit in place of the traditional business plan.

The Lean Startup approach has delivered impressive results, in part because the tools and techniques rely on an entrepreneurial company culture and a customer base of early adopters. This has business leaders asking, "Can our established enterprise act like a startup?" Transferring Lean Startup principles into established enterprises involves significant challenges.

Waverly Deutsch, Clinical Professor of Entrepreneurship at the University of Chicago Booth School of Business, reported[3] that the burning question she gets from CEOs is "How can we be more entrepreneurial?" Her answer is that established organizations can't be fully entrepreneurial, but they can practice experimentation.

Established organizations can't be fully entrepreneurial, but they can practice experimentation.

In established enterprises, product development processes must work within product, market, technology, and business model constraints. Working outside these parameters is difficult because culture, processes and practices do not support it. Startups face very different constraints. Table 8.2 outlines the significant differences we see between these two types of organizations.

Table 8.2. Established Enterprises vs. Startups

Established Enterprises	Startups
Existing infrastructure (sales force, distribution, service, customer base, etc.) imposes constraints on product development	A critical goal is figuring out how to create, sell, distribute, and support the product at an appropriate cost and price
Departments with defined skill sets (marketing, design, engineering, operations, etc.)	Many roles for each person
Policies, procedures, and processes for HR, operations, product development, etc	Policies, procedures, and processes, if any, are typically in flux
Established supplier base	No established suppliers; finding suppliers is part of the task
Defined markets	The focus is on finding the right product for the right market (product-market fit)
Salary with bonuses	Entrepreneur earns money if lucky
Multiple product categories	Typically, one product
Work primarily directed to maintenance and support of existing legacy products	Work devoted to the development of one product with a unique value proposition
Funding from existing profitable products	Funding from investors, including family and friends
An established culture that is difficult to change	Culture is evolving
Typically, the risk is limited to traditional areas: budget, regulatory and technical	Risk is part of the game
Return on investment typically must meet an agreed-upon threshold in the range of hundreds of thousands to millions of dollars	The primary goal is product-market fit with a scalable business model to generate profit

It is challenging for enterprises to adopt startup practices. However, there are other practices that enterprises can adopt that enable greater adaptability and the ability to work outside the usual constraints. One such practice includes internal incubators, which we will discuss in the next section.

Internal Incubators

One way for enterprises to pursue adaptability is through internal incubators. Owens and Fernandez refer to an internal incubator as an Innovation Colony. They recommend creating a special environment within the enterprise, "An environment that provides autonomy, incentive, and focus needed to innovate." Such an environment is separate from but funded by the "Mother Company."[4]

Innovation Colonies have been referred to as Tiger Teams, Autonomous Teams, Internal Incubators, NewCo, and Skunk Works.[5] They have been in existence for some time (Kelly Johnson, an aeronautical and systems engineer of Lockheed Aircraft Corporation, established Skunk Works practices in 1943.[6]). Their specific practices vary somewhat, but they all describe teams operating outside the regular organizational structure to develop a highly innovative or strategically important project. In some instances, the team may be located off-site and separate from the rest of the core enterprise. For simplicity's sake, we will use the term internal incubators to refer to these practices.

Kelly Johnson, an aeronautical and systems engineer of Lockheed Aircraft Corporation, started Skunk Works practices in 1943.

Over the years, we have found trade-offs to using internal incubators. Advantages include speed, project focus, and team camaraderie. Disadvantages include skilled personnel leaving the company once the project is completed, internal employee jealousy, and the "not-invented-here" syndrome. Let's take a closer look at some of these pros and cons.

One positive attribute of internal incubators is speed. One of our clients, a consumer electronics manufacturer, decided to form an internal incubator to develop a product that could displace a major competitor at the Consumer Electronics Show (CES), which was only nine months away. The typical product development project took two to three years. The new product incorporated a licensed technology, which was new to the company adding to the product risk. The project was highly successful, and the team hit the nine-month target while having fun doing it.

What enabled the success of this project? The small, dedicated team (three engineers, one product manager) worked closely together and gave the project their total attention. Compared with the frictions commonly found in the core enterprise, the team structure and culture were very cooperative and collaborative. They developed practices that worked for them, including their working hours, a separate workplace location, and a less constrained dress code.

The merits of this approach are supported by a research study of two separate development groups, the more successful of which worked as a small subgroup focusing on one thing at a time.[7]

A significant disadvantage of the internal incubator approach is that once the product is launched and the group is disbanded, many team members leave the enterprise because they can't fit into the core enterprise culture or the enterprise doesn't know what to do with them.

Another potential side effect is jealousy of non-incubator employees. In a consumer packaged goods (CPG) company, members of the core enterprise viewed the internal incubator group as having all the fun, working on innovative projects. In contrast, those in the core enterprise were stuck with daily mundane tasks. Further aggravating the situation, members of the internal incubator team typically worked and ate lunch separately since the projects they were working on were confidential, making this team look even more privileged.

Yet further problems can occur when those in the core enterprise dislike having an outside team inventing and proposing ideas for them to develop versus having those capabilities built into their own organization. This so-called "not invented here syndrome" plagued Motorola's Early-Stage Accelerator (ESA) program,[8] which was eventually shut down.

Overarching these internal incubator issues is the lack of sustainability and the failure to integrate valuable practices into the core enterprise. After project completion, it is not uncommon for the disbanded team members to leave the enterprise, taking their successful practices, process knowledge, and other expertise with them.

A significant issue with internal incubators is the lack of sustainability.

Govindarajan and Trimble also highlighted the unsustainable nature of the internal incubator model. Their most surprising finding was a lack of learning in teams executing strategic experiments; none of the internal incubators researched implemented a robust learning process.[9]

An ongoing internal incubator program could address sustainability by integrating core employees into these programs on a rotational basis, thereby providing a mechanism to feed learnings back into the entire enterprise. Another approach to increase learning within an enterprise is to foster the principles of a learning organization as we discuss in the next section.

Principles of a Learning Organization

Given that learning is a critical component of ExPD, in this next section, we will review some of the building blocks of a learning organization and how these principles contribute to the infrastructure for ExPD and an adaptive enterprise.

There are numerous definitions of a learning organization, but the explanation that resonates the best for us is, "Organizational Learning is defined as increasing an organization's capacity to take effective action."[10] Identifying and responding to competitive threats and rapidly changing technologies and markets are vital components of building an adaptive enterprise.

As we discussed in Chapter 3, "Why Strategy Matters for Product Development," ExPD emphasizes **exploratory learning,** one of two types of learning:[11]

1. Exploitative learning focuses **internally,** leveraging the existing product/market to improve current technologies, competencies, and business models.

2. Exploratory learning is more **outwardly** focused, using experimentation to discover new technologies, build new competencies, and create new business models. In this form of learning, adaptability is of the utmost importance.

According to Garvin, Edmonson, and Gino,[12] a learning organization has three major building blocks (Figure 8.3): Psychological Safety, Processes & Practices, and Leadership.

Figure 8.3: Learning Organization

- Speak freely
- Take risks
- Time to reflect

Psychological Safety

Processes & Practices

- Experimentation
- Information sharing
- Learning forums

Leadership

- Probing questions
- Listening
- Encourage discussion

1. **Psychological Safety.** People feel psychologically safe if they can speak freely without retractions, appreciate differences in opinions, be open to new ideas, and are encouraged to take risks. Organizations promote psychological safety by valuing and rewarding these behaviors. Psychological safety was listed as the number-one dynamic needed for effective teams at Google (Sidebar 8.1).

Sidebar 8.1. Psychological Safety Supports Teamwork at Google

Google studied 180 teams in its sales and engineering departments and conducted hundreds of interviews to understand and improve teamwork.[13] They found that the attributes of individual members weren't significantly associated with stronger team performance.

What did matter was the dynamics of the team. The researchers concluded that management's emphasis shouldn't be on staffing but on giving team members insight into working together effectively. The analysis showed that the most critical dynamic of the teams was psychological safety. In other words, the team members wanted to feel that they could be safe and that their ideas would be respected and considered, even if they conflicted with the other team members.

2. **Processes and Practices.** A learning organization sets up and maintains processes and practices that promote learning. These processes include ExPD and other exploratory and experimentation methods.

   Experimentation is important. According to the i2020 research study, overperforming organizations are three times as likely as underperformers to embrace a culture of experimentation (40 percent versus 13 percent).[14]

Overperforming organizations are three times as likely as underperformers to embrace a culture of experimentation (40 percent versus 13 percent).

3. **Leadership.** Managers in a learning organization acknowledge their limitations, ask probing questions, listen attentively, and encourage multiple points of view.

Not surprisingly, these three building blocks overlap, reinforcing one another to achieve a learning organization. To check if your organization incorporates these building blocks, see Sidebar 8.2.

Sidebar 8.2. Is Your Organization Functioning as a Learning Organization?

Gavin, Edmonson, and Gino provide a link to a Learning Organization Survey online at https://hbs.qualtrics.com/jfe/form/SV_b7rYZGRxuMEyHRz?Q_JFE=qdg. You can take the survey to determine the extent to which your organization functions as a learning organization and the factors that affect learning within your organization.

The learning organization acts as a catalyst for adaptation and responsiveness through experimentation and learning, and these principles build the foundation for ExPD. However, most organizations have difficulty integrating these principles. As Garvin and his colleagues indicate, most organizations operate by adhering to routines and following strict processes.[15]

Think of the strict rules laid out in a traditional phased-and-gated process with prescribed activities for each phase, even though some of these activities have little or no value. In some instances, important activities that address unique product uncertainties may be overlooked.

ExPD, in contrast, is an adaptable process, requiring project team members to learn through research and experimentation—a vital component of a learning organization.

Management

In this section, we discuss current management practices that work well with the implementation of ExPD. Within this user guide, we define the scope of the manager role to include the organization's future stability, strategy, and product portfolio. Managers are also directly involved with product development and the project teams.

Several management practices will assist a team in becoming more adaptable: decentralized decision-making, including mission-type orders, management by exception, and leading by asking questions.

Decentralized Decision-Making

Adaptability and the ability to react quickly are paramount to successful product development and are the cornerstone of ExPD methodology. What does this mean for management? The best approach is less command and control and more decentralized decision-making. Decentralized decision-making is a mindset shift for some managers because it requires trusting the project team. Some managers want detailed up-front planning, constant updates, and complete control, but this management style does not align well with adaptability.

Adaptability and the ability to react quickly are paramount to successful product development and are the cornerstone of ExPD methodology.

For the team to make good decisions, it must have clearly stated operating parameters. In Chapter 6 (Plan), we refer to these operating parameters as the "Pledge" (Figure 8.4).

Figure 8.4: Pledge Parameters

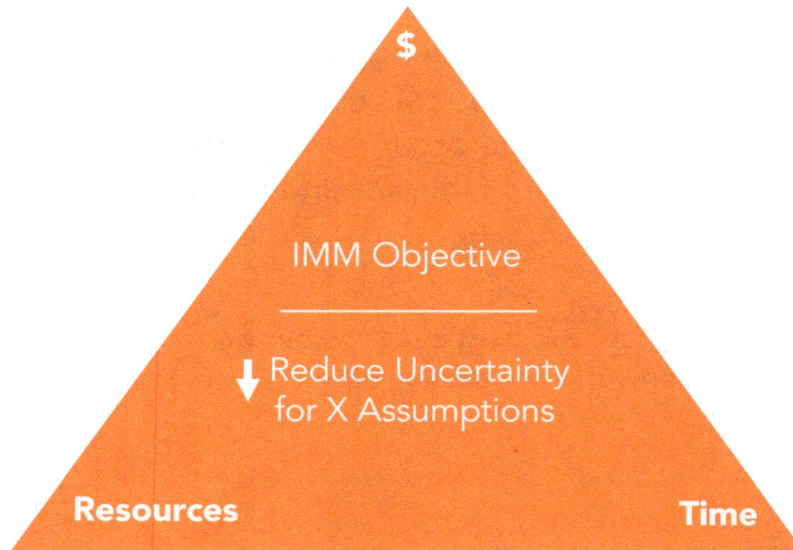

$
\begin{array}{c}
\$ \\
\text{IMM Objective} \\
\hline
\downarrow \text{ Reduce Uncertainty for X Assumptions} \\
\text{Resources} \qquad \text{Time}
\end{array}
$

It is not management's responsibility to tell the team how to proceed with the project at a micro level but rather to provide guidelines that facilitate adaptation to learnings. In a study on formal controls and team adaptability,[16] two fundamental principles have been identified for supporting successful product development teams. The first is target rigidity, where fixed parameters and controls are agreed upon before the project start (similar to the Pledge). The second is process autonomy, which allows flexibility in execution, and adaptation to new project circumstances, including changes in the market or technology.

Target rigidity (parameters set in the Pledge) and process autonomy support team adaptability.

An effective way to enable process autonomy is to issue **mission-type orders.** This practice has been used successfully in the military and can be adapted by organizations. As described in the Maneuver Warfare Handbook, mission-type orders are intrinsic to decentralization.[17]

Figure 8.5: Mission-type Orders

Commander's Intent
Long-term contract - vision and final result

 *Reversed communication flow*

Mission
Short-term contract - react to the immediate situation

Mission-type orders are a contract between superiors and subordinates, and it includes two paths that correspond to the long-term vision and the short-term reaction to an immediate situation (Figure 8.5):

1. The **commander's intent** is the long-term view of what needs to be accomplished. The subordinate needs to understand two levels up, and the contract is to serve the superior's intent on what needs to be performed. In turn, the superior allows the subordinate greater freedom of action regarding how the goal is accomplished.

2. The second contract is the **mission,** which is made on a shorter-term basis. The mission is a "slice" of the commander's intent. The subordinate agrees to support the mission provided they can choose how to achieve it.

In the context of ExPD, the commander's intent and long-term contract are broadly equivalent to the ExPD Product Charter, and represent the high-level vision of the future product and what needs to be accomplished. Meanwhile, the mission and the short-term contract translate to ExPD's Resolve Develop Plan (RDP), which the team executes during the project, as defined in Chapter 6 (Plan).

General Stanley McChrystal further explores the concept of mission-type orders in *Team of Teams: New Rules of Engagement for a Complex World.*[18] In battlefield settings, McChrystal argues that following the conventional 'chain of command' discipline becomes too costly when attempting to fulfill detailed information requests from a subordinate to a distant officer and decreases responsiveness to an agile enemy. A successful military practice includes reversing the communication flow to ensure that the top listens when the bottom speaks.

Reversing the communication flow in project teams make sense. It recognizes that the project team understands the product workings in much greater detail than management, enabling the team to react quickly within the parameters of the Pledge.

A successful military practice includes reversing the communication flow to ensure that the top listens when the bottom speaks.

Management by Exception

Management by exception helps facilitate decentralized decision-making. It is based on a fundamental principle of information theory: project status reports have no value when their findings fall within acceptable margins. The only valuable information is an exception. The main point is that you can virtually eliminate the excessive reporting of routine status updates when you use management by exception. As outlined in the Pledge, the team only meets with management to discuss the status when the project drifts outside acceptable predetermined limits.

Instead of constantly interrupting the project team with project status meetings, it is more efficient to establish boundaries for the team's activities and then require the team to report any exceptions outlined in the Pledge. If the team cannot meet their targets, they need to report to management immediately.

An easy-to-understand application of this principle is statistical process control (SPC), which involves charts such as in Figure 8.6. The team establishes acceptable ranges, which it uses to monitor project behavior. The specified value is the center horizontal line in the SPC chart, and the acceptable range is the area between the top and bottom lines.

Figure 8.6: Understanding Boundaries

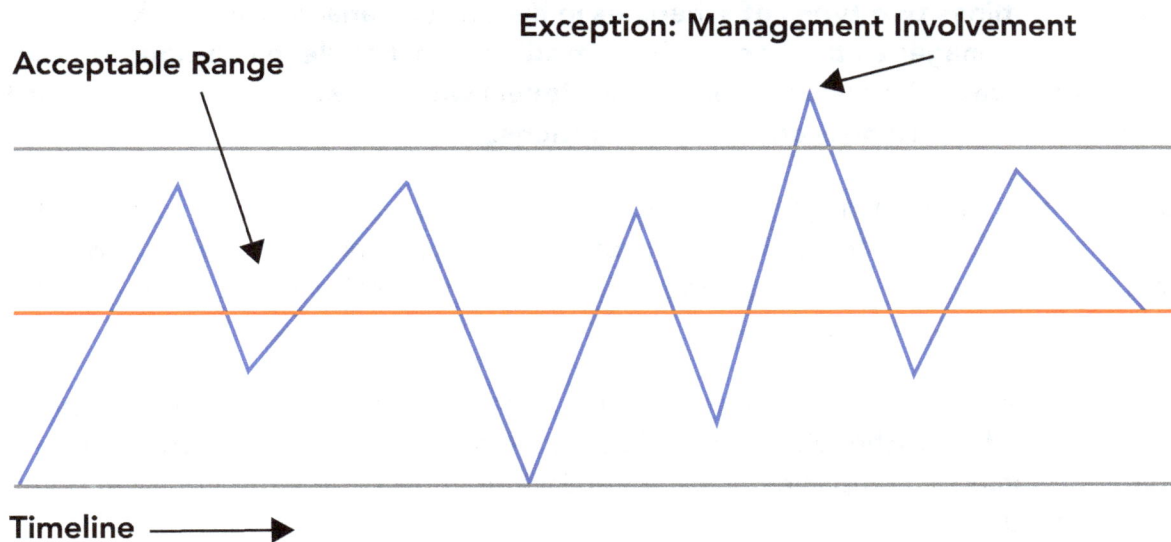

Action is necessary if the process goes above or below the acceptable range, and the project team must inform management. If the measured performance stays within the acceptable range, management does not need to be involved, allowing the team to work without unnecessary distractions and meetings.

Applying this practice to product development projects is a foundation of ExPD and provides your team with the necessary parameters (Pledge) and guidelines for making easy, rapid, and adaptable decisions.

It is also vital that managers respect the schedule of their product developers. Each scheduled meeting represents an exception to the product developer and acts as an interruption that can lead to product delays (Sidebar 8.3).

Sidebar 8.3: Respect the Makers Schedule

Paul Graham outlines two types of schedules in his essay: managers and makers.[19] He defines the manager as the boss with the traditional schedule in half-hour or hour intervals. In contrast, the maker (product developer) works best in half-day intervals to create their product without constant interruptions.

Issues arise when the manager expects the maker to follow the manager's schedule. For a maker, a meeting is like introducing an exception. We have worked in companies where managers bragged about who had the most meetings on their calendars; it was a sign of popularity and perceived value.

Savvy managers know that it is essential to respect the maker's schedule. Graham proposes that a work-around is to schedule meetings later in the day so valuable work isn't interrupted. We have also worked in companies where some days are meeting free, except for important 15-minute project updates.

Leading by Asking Questions

Another helpful practice when managing project teams is to lead by asking questions. The combination of probing questioning and attentive listening reduces the tendency towards command-and-control leadership and promotes decentralized decision-making.

A compelling example of leading by asking questions was demonstrated by a senior product development leader who deliberately avoided micromanaging or telling his team what to do, and instead would ask, "As a team, what do we know and what don't we know about this product?" He led by owning the questions, helping his team understand some of the project's finer points based on his experience of the organization and products. In turn, the team felt trusted, empowered, and free to discuss the nuances of the product issues at hand.

Savvy managers accept the reality that they will confront significant uncertainties and the unexpected in product development. The manager's responsibility is to build the right environment, so the team can efficiently and adequately develop a product. This kind of leadership calls for an honest willingness to discuss with the project team that "I don't have all the answers, so how can we work as a team to navigate the development of this product?"

Team

In this section, we shift our focus from management practices to the product development team itself. In practice, the topics of management and team are intertwined.

While thinking of them separately can be challenging, we can identify several essential teamwork attributes intrinsic to ExPD: cross-collaboration, trust, and employee mastery.

Cross-Collaboration

Although most organizations understand the benefits of cross-functional teams, we still find that most project team members do not collaborate fully. Based on our experience, fiefdoms are more prevalent with companies that use a phased-and-gated process.

Phased-and-gated processes encourage fiefdoms, so disciplines don't mix well.

In a phased-and-gated process, responsibility shifts by phase (Figure 8.7): product management or upstream marketing owns the early phases of the process (Discovery, Scoping, and Business Case); engineering owns Development and Verification & Validation (V&V), and operations and downstream marketing own the Launch phase. This traditional process typically results in ownership issues within the organization, which leads to dysfunction and conflict between disciplines.

Figure 8.7: Typical Process Ownership in a Phased-and-Gated Process

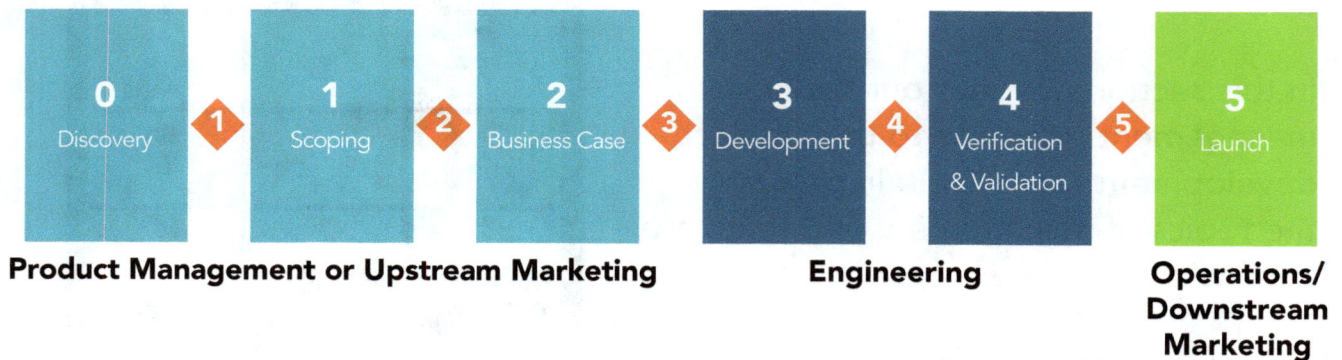

| 0 Discovery | 1 Scoping | 2 Business Case | 3 Development | 4 Verification & Validation | 5 Launch |

Product Management or Upstream Marketing — **Engineering** — **Operations/ Downstream Marketing**

We designed ExPD to infuse cross-collaboration across disciplines to avoid this conflict, starting at the earliest part of the process (Figure 8.8). During Strategy, a cross-functional team is responsible for creating enterprise, business, and product direction. Ideas & Selection includes the use of cross-functional teams in determining the best product ideas. During Investigate, the project team starts with the chosen idea and works cross-functionally to identify, evaluate, and prioritize product uncertainties. Finally, during Explore & Create, the functions work either in parallel or cross-functionally to resolve the most critical uncertainties through experimentation and research.

Figure 8.8. Collaboration within the ExPD Process

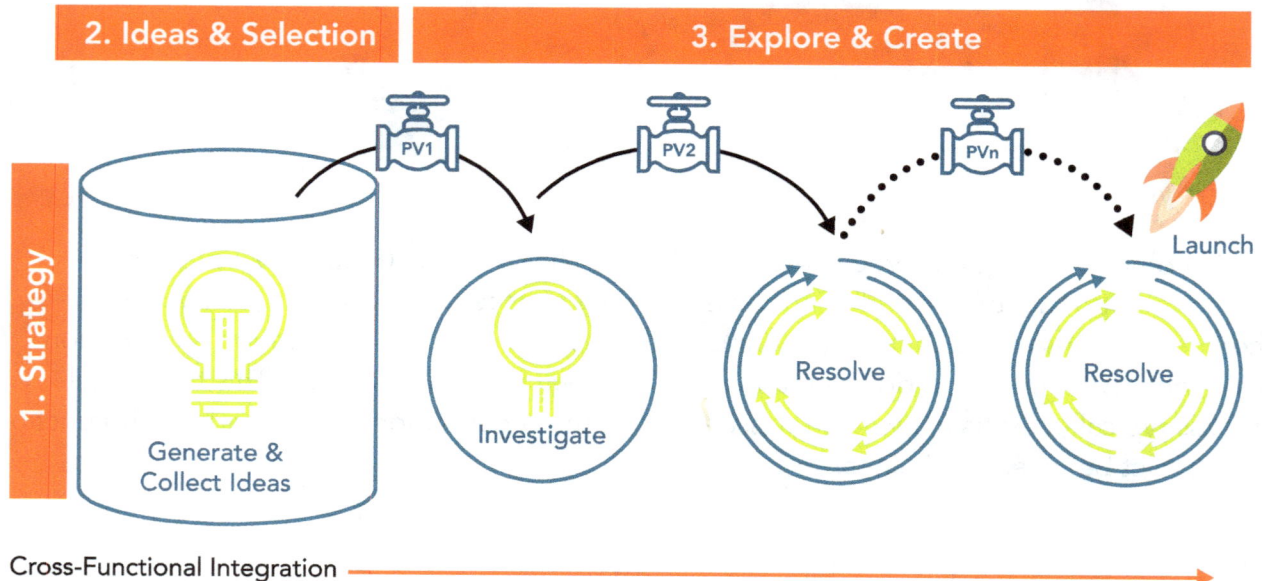

2. Ideas & Selection   3. Explore & Create

1. Strategy

Generate & Collect Ideas — Investigate — Resolve — Resolve — Launch

Cross-Functional Integration

In ExPD, cross-collaboration is essential, which may be a sticking point for some organizations. This requires a conscious effort from management in redesigning roles and responsibilities and the team structure. As discussed in the next section, a key component of successful cross-functional teams is the trust and respect required of all parties.

Building Team Trust

In *Overcoming the Five Dysfunctions of a Team,* **Patrick Lencioni highlights the importance of vulnerability when building trust. Vulnerability-based trust is based on team members being unafraid to admit their mistakes and choosing not to participate in political behavior.**[20]

Let's be candid; most team members don't want to highlight their shortcomings since it can affect their performance review. As far as not partaking in political behavior, unfortunately, most people like to gossip.

Trust is one of the most challenging qualities to instill within your organization. As consultants, we have seen both ends of the spectrum, from a very trusting team to belligerent, untrusting team members. Mistrust is especially evident when the different disciplines don't value each other's participation, contributions, or role in the project.

Despite the difficulties, organizational initiatives can improve trust between individual team members and across disciplines. Achieving the following practices can improve trust:

- Team members:
    - o understand their roles and what they bring to the project
    - o feel valued, engaged, and listened to
    - o can depend on their coworkers to deliver
    - o share input without fear of retribution
    - o feel psychological safety and know their ideas will be respected (Sidebar 8.1)

- Teams are focused, and preferably small

- Team membership is consistent, versus constant turnover

- Face-to-face communication (even if it is on Zoom) is frequent

- Project findings and information are transparent

- Team meetings reinforce learning instead of blame or failure

As a tool for maximizing communication, trust, and adaptability, we recommend retrospective meetings. A retrospective is "a special meeting where the team gathers after completing an increment of work to inspect and adapt their methods and teamwork."[21] In ExPD, the primary purpose is to consider what was learned and adapt the project accordingly. Retrospectives are not to be confused with post-mortem or post-launch meetings which occur at the completion of the project.

> *Retrospectives are not to be confused with post-mortem or post-launch meetings which occur at the completion of the project.*

Retrospectives help build team trust by highlighting possible team issues, which is much more effective than letting team issues fester throughout the project. During a retrospective, you can get real-time team results about team satisfaction by collecting team member ratings on a scale from 1 (dissatisfied with the team) to 5 (delighted with the team).[22]

Retrospectives can be facilitated as a 15-minute stand-up meeting. Still, they should run as long as necessary to achieve the objective of helping the team adapt in real-time to new project circumstances. Meeting frequency depends upon the project's urgency and the speed at which the assumptions are resolved. Typically, the team leader asks the following questions:

- What progress have you made since the last meeting?

- How will you work toward your next key milestone?

- Do you face any issues or obstacles?

Besides assessing project status, you should also include the following questions to ensure the team is in alignment:

- What assumptions have been resolved?

- Where is the project in the Idea Maturity Model?

- Are we on track with the Product Charter and RDP?

- Are there any dealbreakers?

- Are we still within the parameters of our Pledge?

- Should we proceed with this project?

These questions help the team determine the next steps and whether anything should be brought to management's attention. For maximum performance and effective communication, we prefer co-located teams. Still, this approach has become more challenging, with team members dispersed across the country or around the globe, as well as physically distanced in response to COVID-19. Even a virtual retrospective is helpful; it ensures that the team communicates, and stays focused on essential project objectives. We have found that retrospectives improve trust, communication, and project performance, especially with virtual teams clearly spelling out what needs to be executed on the project.

Research has shown that the overall effect of working remotely is not necessarily detrimental to the team's performance. In fact, dispersed teams can outperform co-located teams if they have the right processes, understand tasks, and clearly understand their contribution to the team.[23]

Research has shown that the overall effect of working remotely is not necessarily detrimental to the team's performance.

Employee Mastery

Working on an ExPD project requires a certain amount of mastery from each team member. We have seen ExPD deliver benefits in organizations where employees are highly skilled and able to work without a high level of structure.

The agile community defined three employee mastery levels:[24]

- **Level 1 (Doing).** Employees at Level 1 need a process to follow. They may be new to the discipline or position

- **Level 2 (Understanding).** At Level 2, employees are open to the possibility of alternative approaches, but they still like having a defined process to fall back on

- **Level 3 (Fluent).** A Level 3 employee is able and willing to adjust and improvise without a specified set of process activities and direction

In general, ExPD is optimized when the team members have a high level of mastery (Level 3).

Level 3 employees are positioned for leading ExPD teams to develop new-to-the-company and riskier products. We highly recommend that Level 1 and 2 employees be part of the ExPD team to improve their mastery level and to be more comfortable with ambiguity when developing products.

To learn about star performers (similar to Level 3 employees) at AT&T Bell Laboratories, read Sidebar 8.4.

Sidebar 8.4. Star Performers at AT&T Bell Laboratories

Seven years of extensive research into the practices and performance of scientists and engineers at AT&T Bell Laboratories found that the vast majority are solid middle performers.[25] However, another 10 to 15 percent of scientists and engineers are star performers.

What makes up a star performer? You may think that innate ability (e.g., IQ tests, personality inventories) is the primary factor in being a star performer. Instead, the researchers found that **nine work strategies** make star performers, listed in the order of importance:

1. **Taking initiative.** Accept responsibility beyond their job titles.

2. **Networking.** Have access to coworkers with specific expertise and share their knowledge with those who need it.

3. **Self-management.** Regulate their performance level and career growth.

4. **Teamwork effectiveness.** Assume joint responsibility for work activities and accomplishing shared goals with coworkers.

5. **Leadership.** Build consensus on common goals and how to achieve them.

6. **Followership.** Help leadership accomplish the organization's goals and don't rely on managerial direction.

7. **Perspective.** Take on other viewpoints, including the customer, manager, and team.

8. **Show-and-tell.** Present their ideas persuasively.

9. **Organizational savvy.** Navigate competing interests in their organization, promote cooperation, address conflict, and get things done.

Interestingly, the researchers found that these work strategies were trainable, and AT&T Bell Labs established a productivity training program. It was based on each of the nine work strategies with internal engineers facilitating these sessions.

Culture

An organization's culture is critical to success but also challenging to see or measure. Merriam-Webster chose the word culture as their Word of the Year in 2014 when it was the word that saw the most significant spike in lookups on their website.[26] People are desperate to understand the word.

People are desperate to understand the word culture.

Culture is a complex and multi-dimensional concept. In fact, Merriam-Webster has six definitions, applying them to disciplines ranging from anthropology and the arts to agriculture and biology.

For ExPD, having the right organizational culture is essential. Elliott Jaques, who initially documented the concept of organizational culture in 1952, defined it as "encompassing values and behaviors that contribute to the unique social and psychological environment of a business."[27] This definition seems too obtuse, so we will refer to a framework we developed and named the Culture Triangle.

Culture Triangle

Earlier in this chapter, we discussed the importance of management instilling a culture of honesty and trust. Instilling a good culture is challenging, but it is easier if you have the correct management, team, and process—the three outer segments of our Culture Triangle (Figure 8.9).

Figure 8.9: Culture Triangle

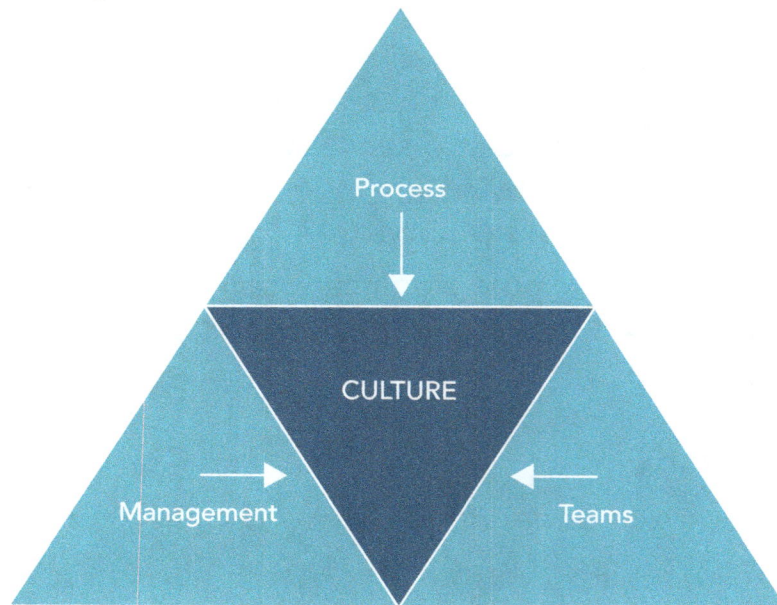

The Culture Triangle illustrates that management, team, and process shape culture, not the other way around. This idea builds on a Harvard Business Review article on organizational culture and performance that claims culture is not the root problem. Instead, the reverse is happening; broken business processes and practices lead to cultural dysfunction. Once these areas are reworked and improved, a high-functioning culture can fall into place.[28]

Broken business processes and practices lead to cultural dysfunction.

We agree with this premise, mainly based on our years of experience working with many organizations. We have seen cultures negatively affected by poor communication, vague or conflicting roles and responsibilities, lack of trust, poorly designed processes, and other management missteps. The following case study illustrates how we used the three outer segments of the Culture Triangle to help transform one company's culture by improving critical processes and practices.

Cultural Change Case Study

The best way to describe this company's culture is conflict and mistrust between management and the project teams. Management and the project teams held monthly project update meetings that lasted a half-day and went into excruciating detail on each project.

When our consultants arrived, the head of engineering refused to attend any project update meetings. We certainly didn't blame him for his reaction since the discussions didn't accomplish anything except for people expressing their anger over project delays, which they blamed on his department.

Management

We had strong backing from management since they were at the end of their rope. Management recognized the need for improving the culture and was willing to invest the time and expense to make the necessary changes. The COO was instrumental in ensuring that the project team was on board with the suggested changes. He worked directly with the project team to design its future state and processes. This commitment level communicated to the team that management was serious about change.

Process

Based on our observations of the monthly reviews, we felt that a time-to-market initiative would reduce conflict within the company, and we targeted this problem area first. We determined the optimal number of projects based on available resources and why products weren't getting to market: 42 projects were currently being worked on, but the company had only enough resources to work on 17.

Once these projects were prioritized and scaled back in the Prioritization Valve, the impact was immediate. Instead of squeezing just one more project into the process, the management team followed the rules and guidelines for prioritization and selection. For the first time, the team started to deliver products on time.

Team

In the area of teamwork, we undertook several initiatives. First, we provided advice on the future organizational structure and the missing essential skills and experience. The company started to hire additional staff, and the project team members felt a lot less pressure to work on multiple projects. The company employed professional project managers to help manage the projects instead of assigning the work to any available engineer.

Also, the teams were designed to be cross-functional, so decisions weren't made in a vacuum. Improved communication between the disciplines positivity affected planning and time to market.

In addition, we instituted 15-minute stand-up meetings, enabling rapid decision-making based on the day's learnings and outcomes. If the team went beyond the Pledge parameters, they escalated this to management.

Finally, project status reporting is based on fact rather than conjecture. Management is now kept up to date on project status via a real-time project status update. The monthly project updates that had been so dysfunctional are now shorter and more efficient since they only report the project exceptions that were not meeting the Pledge parameters.

Project status reporting is based on fact rather than conjecture.

Is the company's culture perfect now? No, but it is a lot better. Employees are more relaxed, and they can communicate without all the drama. Still, senior management occasionally wants to squeeze another project into the pipeline ("It's tiny; can't we just do this one?").

It isn't easy to improve culture, and there are probably other variables that need to be considered based on unique organizational characteristics. But a good starting point for changing culture is to think about the Culture Triangle variables (management, process, and team) and how they can be improved within an organization.

Learning from Failure

A cultural value crucial for innovating—with or without ExPD—is acceptance of failure. Some organizations embrace and celebrate failure. Businesses in the startup community have FailCon conferences to study each other's failures. NASA has a Lean Forward; Fail Smart Award, and Tata Group has an award it calls "Dare to Try."[29] Leaders employing ExPD can remind their employees, "The faster we learn and adapt, the faster we succeed."

"The faster we learn and adapt, the faster we succeed."

Some enterprises have difficulty embracing a mindset of learning and adapting since they have long based their management processes and risk controls on predictability and control. So much is at stake to do well and be associated with the "winning project." Even when top managers voice the importance of risk-taking, employees do everything to avoid it. They are afraid they will be blamed for any loss and subsequently lose their job.

Management must encourage teams to continuously learn and frequently update their decisions and plans based on the current findings and data. The Turba team pivoted fast and adapted the PERS product, as described in Chapter 7, "Resolve," Sidebar 7.7. The unexpected competition from JinQ did not substantially set the Turba team back. They reviewed the latest competitive findings and leveraged the business model, product strategies, and product roadmap to quickly pivot to the Max Model without significantly impacting budget, resources, or time.

The idea of learning fast and adapting is a significant mind shift for many organizations. During our NSF research, some product leaders in large enterprises were uncomfortable addressing product risks. Why is there an adverse reaction to identifying risks when a project is initiated? The answer commonly comes back to culture. In large enterprises, employees proposing a product idea want it to succeed and don't want to show any potential weaknesses or failures.

According to Ross, Beath, and Mocker, this mindset change will entail upending established management practices and forcing a change in corporate culture. Embracing failure is foreign to most product leaders who have risen in large enterprises today. Companies may need to call upon the next generation of product leaders to change to an iterative test-and-learn culture that enables learning and adapting.[30]

Learning and adapting are foreign to most product leaders who have risen in large enterprises today.

Suppose enterprises don't have the time to wait for the next generation? In this case, the organization must fix critical processes and practices, as outlined in the Culture Triangle (Figure 8.9).

This change, like any organizational transformation is hard, and ExPD will help since it provides an approach to operating in a rapidly evolving environment. And as described in this chapter, ExPD works best in an adaptable enterprise characterized by management approaches that decentralize decision-making; cross-functional teamwork built on trust; and a culture that embraces learning and adapting.

Key Chapter Points

1. Today, most enterprises are in an uncertain environment, so it is vital to create an internal environment with dynamic capabilities to adapt to uncertainties and risks.

2. An enterprise is a permanent organization designed to execute a repeatable and scalable business model, while a startup is a temporary organization that looks for a repeatable and scalable business model.

3. An alternative for large enterprises that want to be more entrepreneurial is experimentation.

4. One way enterprises have pursued adaptability is through internal incubators, which have had mixed success and have generally proven unsustainable.

5. Vital components of building an adaptive enterprise include acting and responding to competitive threats and rapidly changing technologies and markets.

6. ExPD is not based on following a cookbook formula for solving problems. Instead, it is an adaptable process that reduces the most critical uncertainties, requiring project team members to think and solve problems through experimentation—a vital component of a learning organization.

7. In a study of new product development projects, target rigidity (provided by the Pledge in ExPD) and process autonomy supported team adaptability.

8. The conventional chain of command constrains the pace and ability to react quickly. Organizations can correct this by reversing the communication flow to ensure that the top listens when the bottom speaks.

9. Instead of constantly interrupting the project team with project status meetings, it is more efficient to establish boundaries for the team's activities and then require the team to report any exceptions.

10. We recommend that the team execute project updates through retrospective meetings to inspect and adapt their methods after completing each major increment of work.

11. Research has shown that working remotely is not necessarily detrimental to the team's performance.

12. Determining the correct employee mastery is a crucial component of ExPD. Level 3 mastery (Fluent) is ideal for leading an ExPD team.

13. Culture is not the root problem when performance is poor. Instead, the reverse is happening: broken business processes and practices lead to cultural dysfunction.

14. Even when top managers voice the importance of risk-taking, employees do everything to avoid it due to the influence of culture, processes, and incentives.

Notes

1. See Chapter 3, *Why Strategy Matters For Product Development*, Table 3.1 "The Characteristics of the Three Types of Environment".

2. Steve Blank, "Why Companies Are Not Startups," personal blog, March 4, 2014, http://steveblank.com/2014/03/04/why-companies-are-not-startups/.

3. The University of Chicago, Booth Women Connect Conference, October 23, 2015, http://www.chicagobooth.edu/faculty/directory/d/waverly-deutsch.

4. Trevor Owens and Obie Fernandez, *The Lean Enterprise: How Corporations Can Innovate Like Startups* (Hoboken, NJ: Wiley, 2014).

5. Ben R. Rish and Leo Janos, Skunk Works: *A Personal Memoir of My Years at Lockheed* (Boston: Little, Brown and Company, 1994). Lockheed Martin coined the term Skunk Works and used these practices starting in 1943.

6. Rish, *Skunk Works*.

7. Rob Cross, Thomas H. Davenport, and Peter Gray, "Collaborate Smarter Not Harder," *MIT Sloan Management Review* (Fall 2019).

8. Mary Drotar, "How Motorola Uses an Early-Stage Accelerator," interview of Jim O'Connor, *Visions* (Product Development & Management Association) 30(5) (December 2006).

9. Vijay Govindarajan and Chris Trimble, *10 Rules for Strategic Innovators: From Idea to Execution* (Boston: Harvard Business School Press, 2005). This provides the ten characteristics of strategic experiments, ranging from high revenue growth potential to testing an unproven business model.

10. Daniel H. Kim, "The Link between Individual and Organizational Learning," *Sloan Management Review* (Fall 1993): 43.

11. Aelong Wei, Yaqun Yi, and Hai Guo, "Organizational Learning Ambidexterity, Strategic Flexibility, and New Product Development," *Journal of Product Innovation Management* 31 (2014): 832–34.

12. David A. Garvin, Amy C. Edmonson, and Francesca Gino: "Is Yours a Learning Organization?" *Harvard Business Review* (March 2008).

13. Leigh Buchanan, "The Most Productive Teams at Google Have These 5 Dynamics." *INC Magazine* (April 12, 2016).

14. Frank van den Driest, Stan Sthanunathan, and Keith Weed, "Building an Insight Engine," *Harvard Business Review* (September 2016), https://hbr.org/2016/09/building-an-insights-engine.

15. Garvin, "Is Yours a Learning Organization?"

16. Nina Detzen, Frank H. M. Verbeeten, Nils Gamm, and Klaus Möller, "Formal Controls and Team Adaptability in New Product Development Projects," *Management Decision* 56(7) (2018): 1541–58.

17. These decentralized practices refer to the Boyd cycle, which is based on maneuver warfare and the OODA (Observe, Orient, Decide, Act) loop. See William Lind, *Maneuver Warfare Handbook* (Routledge; August 6, 1985).

18. General Stanley McChrystal with Tantum Collins, David Silverman, and Chris Fussell, *Team of Teams: New Rules of Engagement for a Complex World* (New York: Penguin, 2015).

19. Paul Graham, Maker's Schedule, Manager's Schedule http://www.paulgraham.com (July 2009).

20. Patrick Lencioni, *Overcoming the Five Dysfunctions of a Team* (San Francisco: Jossey-Bass, 2005).

21. Esther Derby and Diana Larsen, Agile Retrospectives: *Making Good Teams Great* (Dallas: Pragmatic Programmers, 2006).

22. Derby, *Agile Retrospectives*.

23. Frank Siebdrat, Martin Hoegl, and Holger Ernst, "How to Manage Virtual Teams," *MIT Sloan Management Review* 50(4) (Summer 2009).

24. Preston Smith, *Flexible Product Development: Building Agility for Changing Markets* (San Francisco: Jossey-Bass, 2007).

25. Robert Kelley and Janet Caplan, "How Bell Labs Creates Star Performers," *Harvard Business Review*, October 7, 2014.

26. Joshua Rothman, "The Meaning of Culture," *New Yorker*, December 26, 2014.

27. Elliott Jaques, *The Changing Culture of a Factory*, 1st American ed. (New York: Dryden Press, 1952).

28. Jay W. Lorsch and Emily McTague, "Culture Is Not the Culprit," *Harvard Business Review* (April 2016).

29. Julian Birkinshaw and Martine Haas, "Increase Your Return on Failure," *Harvard Business Review* (May 2016).

30. Jeanne W. Ross, Cynthia M. Beath, and Martin Mocker, "Creating Digital Offerings Customers Will Buy," *MIT Sloan Management Review* (Fall 2019

Next Steps

We provided the necessary guidance and practices to implement the ExPD process within this user guide. It was enhanced with the Turba case studies that demonstrated how ExPD was executed starting with the first segment, "Strategy," through the final segment, "Explore and Create."

If you're serious about getting started with the ExPD process, we recommend the following next steps:

- Pilot the process with a product idea that has a medium to a high level of uncertainty attached to it

- Follow the ExPD methodology as outlined in this user guide, at the highest level:

    o Assign a cross-functional team to identify, evaluate and prioritize the most impactful product uncertainties and risks

    o Ensure the team understands the business model, product strategies, and roadmap surrounding the product idea

    o Determine the maturity of the product idea using the Idea Maturity Model (IMM)

    o Plan the project using the Product Charter and Resolve Development Plan (RDP)

    o Resolve the most impactful uncertainties through experimentation and research

    o Learn and adapt from the most recent data and findings

- Ensure that you have the correct infrastructure (people, practices, and tools) to support the project

Some enterprises may still prefer the traditional phased-and-gated approach since the activities are easy for most employees to understand. A phased-and-gated process is also useful when making a simple product revision with minimal uncertainty. Unfortunately, we have experienced simple revisions that keep the project team spinning because of unforeseen uncertainties and risks, especially in the Develop phase, where technology risks are high. If your organization wants to use the phased-and-gated approach (Figure A), we highly recommend integrating assumptions within the process. However, this is less efficient and more rigid than a complete ExPD system, and it adds some additional work to the phased-and-gated process.

Figure A: Assumptions Integrated with the Phased-and-Gated Process

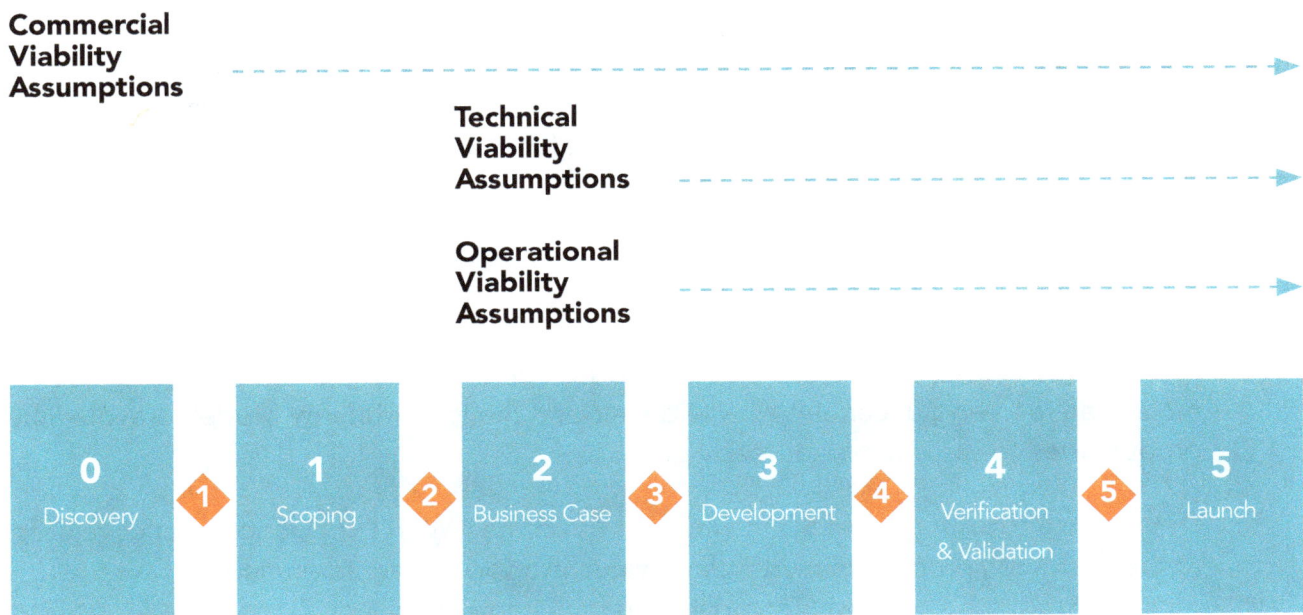

Integrating assumptions into the phased-and-gated process is a way to test whether ExPD principles fit your organization, but you will not realize all the benefits, especially adaptability.

Closing

Implementing the entire ExPD process will enable your enterprise to realize the following benefits:

1. **Speed.** ExPD accelerates the product development process in multiple ways, including removing bureaucracy, reducing batches of work, eliminating rework, and focusing on the most critical product uncertainties.

2. **Fail Fast and Early.** ExPD identifies, evaluates, and prioritizes uncertainties early in the process to detect and adapt to the unique nuances of the product. This enables the project team to learn quickly, and if necessary, fail fast and early.

3. **Achieving Strategic Alignment.** A strategy is critical to product development. ExPD incorporates the s2m Strategic Framework™, a powerful tool that adapts to different markets and environments.

4. **Launching a Market-Ready Product.** ExPD recognizes that product development is more than resolving uncertainties. We developed the Idea Maturity Model to evolve the product idea towards a market-ready product.

5. **Developing Products That Start with the Customer.** With ExPD, uncertainties related to the customer, product-market fit, and value proposition are identified early in the process. Products also fit with company strategy and capabilities.

6. **Prioritizing and Optimizing Projects Real-time.** We developed prioritization valves within the ExPD process to ensure more timely resourcing. The pipeline is easily overloaded without the valves and can slow to a crawl.

7. **Learning and Adapting.** With uncertainty, the best outcomes are achieved through experimentation and research. Regardless of the outcome, learnings are of paramount importance and adapting to unique product nuances is imperative.

8. **Decreasing Unnecessary Activities.** Activities and documentation are often performed mechanically within the traditional phased-and-gated process. In contrast, ExPD customizes activities and documentation based on the unique nuances and uncertainty of the product. This may mean skipping some activities or adding new activities to accommodate product nuances.

9. **Improving Decision-making.** ExPD is based on decentralized decision-making. To make high-quality decisions, clearly stated parameters (Pledge) are developed, making it easy for the team to understand what needs to be completed on the project without constant interruptions.

10. **Reducing Product Development Cost.** The 9-benefits listed above lead to the reduction of product development costs.

Enjoy discovering the unknown.

Mary & Kathy

About the Authors:

Mary and Kathy founded the product development firm Strategy 2 Market® (s2m) in 2002. s2m specializes in helping large and mid-sized enterprises improve their product development processes.

They went on a mission to improve the traditional phased-and-gated product development process that most enterprises use today. They developed an alternative adaptable process called Exploratory PD® (ExPD) that ultimately drives down and manages the most critical product development uncertainties and risks. Out of this process, they developed the Product Risk Framework® tool.

They received a grant in 2018 from the NSF STEM I-Corp program sponsored by the University of Chicago Polsky Center for Entrepreneurship and Innovation. The grant was for the Product Risk Framework (PRF), which supports the ExPD process. Mary and Kathy are both graduates of the Booth School of Business at the University of Chicago.

Mary Drotar
Co-Founder, Strategy 2 Market

in linkedin.com/in/marydrotar

Kathy Morrissey
Co-Founder, Strategy 2 Market

in linkedin.com/in/kathymorrissey

Index